T0180225

# Can Regional Integration Arrangements Enforce Trade Discipline? The Story of the EU Enlargement

*Also by Zdenek Drabek*

GLOBALIZATION UNDER THREAT: The Stability of Trade Policy and Multilateral Agreements

FINANCIAL REFORM IN CENTRAL AND EASTERN EUROPE
(*co-edited with Stephany Griffith-Jones*)

# Can Regional Integration Arrangements Enforce Trade Discipline?

## The Story of the EU Enlargement

Edited by

Zdenek Drabek
*World Trade Organization*
*Switzerland*

First published 2005 by
PALGRAVE MACMILLAN
Houndmills, Basingstoke, Hampshire RG21 6XS and
175 Fifth Avenue, New York, N.Y. 10010
Companies and representatives throughout the world

PALGRAVE MACMILLAN is the global academic imprint of the Palgrave Macmillan division of St. Martin's Press, LLC and of Palgrave Macmillan Ltd. Macmillan® is a registered trademark in the United States, United Kingdom and other countries. Palgrave is a registered trademark in the European Union and other countries.

ISBN 978-1-349-72847-3

This book is printed on paper suitable for recycling and made from fully managed and sustained forest sources.

A catalogue record for this book is available from the British Library.

Library of Congress Cataloging-in-Publication Data
Can regional integration arrangements enforce trade discipline? : the story of the EU enlargement / edited by Zdenek Drabek
    p. cm.
Includes bibliographical references and index.

1. Free trade–Government policy–European Union countries. 2. European Union countries–Commercial policy. 3. Regionalism–European Union countries. 4. Free trade–Government policy–Europe, Eastern. 5. Europe, Eastern–Commercial policy. 6. Regionalism–Europe, Eastern. 7. Free trade–Government policy–Europe, Central. 8. Europe, Central–Commercial policy. 9. Regionalism–Europe, Central. I. Drabek, Zdenek.
HF2036.C36 2005
382'.9142–dc22                                                    2004054252

10   9   8   7   6   5   4   3   2   1
14   13   12   11   10   09   08   07   06   05

# Contents

# List of Figures

# List of Tables

# Acknowledgements

Most of the contributions to the book result from a research project 'The Stability of Trade Policy and the Accession of Central and East European Countries into the European Union'. The papers by Cernat and Laird and by Drabek were written subsequently, after the conclusion of the project. This research was undertaken with support from the European Union's Phare ACE Programme 1997. The content of the publication is the sole responsibility of the author and it in no way represents the views of the Commission or its services. The views also do not necessarily represent the views of the World Trade Organization (WTO) Members or the Secretariat. The generous assistance of the European Commission is gratefully acknowledged.

The project and the individual authors benefited from rich discussions of the papers at a seminar that took place in Prague on May 26–27, 2000. The seminar was organized in collaboration with the Institute of Economic Studies, under the capable management of professor Michal Mejstrik, and with the Faculty of Social Studies of Charles University, headed by Professor Ludek Mlcoch, Dean of Social Sciences. The seminar was organized under the auspices of the Deputy Minister of Finance and of the Governor of the Central Bank of the Czech Republic. The meeting provided an excellent opportunity for the authors of the papers to discuss their drafts both among themselves and with invited participants to the seminars. The editor is particularly grateful to Vladimir Dlouhy (former Minister of Trade and Industry of the Czech Republic), Karel Dyba (former Minister of Economy of the Czech Republic), Zdenek Tuma (Governor of the Czech National Bank), Jan Mladek (Member of Parliament and a former Deputy Minister of Finance) for their support of the project and for their active participation at the seminar. The project also benefited from two public lectures given by Professors David Begg and Alasdair Smith. Unfortunately, the latter, who was an original member of the research team, was unable to finalize his contribution for this volume as he had to take up the important post of Vice Chancellor of Sussex University.

Additional useful comments were provided on various aspects of the WTO Agreements, NAFTA and the European Union by WTO staff. I am grateful to my colleagues Robert Anderson, Jan Bohonos, Richard Eglin, John Finn, Mario Kakabadze, Eki Kim, Vasile Kulakoglu, Vivien Liu and Gretchen Stanton. My thanks also go to Ganzorig Arslanbaatar, my Mongolian assistant, who spent a short period of time at the WTO as an intern but whose intelligence and hard work were brilliant and way beyond his assignments. No person mentioned here should any way be held responsible for any misinterpretations, mistakes or other shortcomings of this book.

I am also very grateful to my wife for her spiritual and psychological support to this project. More importantly, her practical assistance and contribution as an editor, administrator and manager was crucial since my own contribution could only be done in my spare time. But neither she nor I would have been able to do our job properly without the support of our children – Bianca and Alexandre – who did not demand 100 per cent of our attention and our time, even though they would have been more than justified to do so.

ZDENEK DRABEK

# Notes on the Contributors

**David Begg** is Principal of Tanaka Business School of Imperial College, London. Until his most recent appointment at the Business School, he was a Full Professor of Economics at Birnbeck College, University of London, having previously taught at the University of Oxford. He has been a visiting scholar at MIT, Princeton, adviser to the Bank of England, the International Monetary Fund and Research Director at the London Business School. He is the Founding Managing Director of *Economic Policy*, the Roll Committee on the Independence of the Bank of England and is author of numerous textbooks and articles mainly in the area of macroeconomics.

**Lucian Cernat** is Economic Affairs Officer with the United Nations Conference on Trade and Development. He has a PhD in economics from the University of Manchester and the MA from the University of Oxford. Prior to Joining UNCTAD, he was a trade diplomat with the Romanian Ministry of Foreign Affairs and a lecturer at University of Bucharest under the Yale – CEP Eastern Scholar Program.

**Zdenek Drabek** is Senior Adviser at the World Trade Organization. He was a Minister's Plenipotentiary in the Czechoslovak and Czech governments and Chief International Negotiator. He worked as economist in the World Bank and was a chairman of the Department of Economics at the University of Buckingham (UK) and Research Officer at the University of Oxford. He has lectured widely in North America, Europe, Japan and East Asia. He was a member of international task forces on *Vietnam*, on *Managing Commodity Risks*, on *Globalization and Poverty Elimination* and of inter-parliamentary conferences on Ukraine and various other countries acceding to the WTO. He has published extensively in the area of international and macro economics.

**Michaela Erbenova** is Chief Executive Director and Member of the Monetary Board of the Czech National Bank (central bank). She was Director of Strategic Planning Department in Komercni Banka prior to her appointment to CNB and chief adviser to the Minister of Finance of the Czech Republic and adviser to the prime minister. She was educated at Charles University in Prague and Moscow State University.

**Tomas Holub** is Adviser to the Governor of the Czech National Bank and Executive Director of its Monetary and Statistics Department. He is a lecturer at Charles University. He was educated at Charles University and the London School of Economics and is a winner of the prestigious Bolzano Prize (Charles University) and Olga Radzyner Prize (Austrian National Bank).

**Miroslav Hrnčír** is Director of Research in the Czech National Bank. He was previously a Member of the Monetary Board in the Czech National Bank and Senior Research Officer at the Economic Institute of the Czechoslovak Academy of Sciences. He is a leading Czech international economist and has published widely on trade policy, exchange rates and macroeconomic policy both in and outside the Czech Republic.

**Sam Laird** is a manager of the Trade Policy Unit at UNCTAD. He was previously employed as Senior Economist at the World Trade Organization and in the Research Department of the World Bank. He is a Special Professor in International Economics at the Leverhulme Centre of the University of Nottingham. He started his career in the Trade Commission in Australia. He has published extensively on trade and trade policy and acted as adviser to many governments in developing countries.

**Sandor Meisel** is Senior Fellow at the Institute for World Economics of the Hungarian Academy of Sciences. He was a member of Hungarian negotiating team on the accession of Hungary to the European Union. Formerly he was a Director in Ministry of International Economic Relations.

**Jan Michalek** is a former Deputy Permanent Representative of Poland to the World Trade Organization and a former Vice-Dean of the Department of Economics at Warsaw University. He is currently Associate Professor in International Economics at Warsaw University. He is a resident scholar in several Western universities and published widely on trade in Western and Polish journals.

# List of Abbreviations

| | |
|---|---|
| APEC | Asia-Pacific Economic Cooperation |
| ASEAN | Association of Southeast Asian Nations |
| ATC | Agreement on Textiles and Clothing |
| CAP | Common Agriculture Policy |
| CCFTA | Canada–Chile Free Trade Agreement |
| CEB | Czech Export Bank |
| CEFTA | Central European Free Trade Agreement |
| CER | Closer Economic Relations (also ANCERTA) |
| CIS | Commonwealth of Independent States |
| CN | Combined Nomenclature |
| CNB | Czech National Bank |
| COMESA | Common Market for Eastern and Southern Africa |
| EA | European Agreement |
| EC | European Communities |
| ECOWAS | Economic Community of West African States |
| EFTA | European Free Trade Association |
| EGAP | Export Guarantee and Insurance Corporation |
| EMU | European Monetary Union |
| EU | European Union |
| FDI | Foreign Direct Investment |
| FTAs | Free Trade Agreements |
| FTAA | Free Trade Area of the Americas |
| GATS | General Agreement on Trade in Services |
| GATT | General Agreement on Tariffs and Trade |
| GSP | Generalized System of Preferences |
| LDCs | Least Developing countries |
| MERCOSUR | Southern Common Market |
| MFN | Most Favoured Nation |
| MRAs | Mutual Recognition Agreements |
| NAFTA | North American Free Trade Agreement |
| NTMs | Non-Tariff Measures |
| OECD | Organisation for Economic Co-operation and Development |
| RIAs | Regional Integration Agreements |
| ROO | Rules of Origin |
| SAARC | South Asian Association for Regional Cooperation |
| SPS | Sanitary and Phytosanitary Measures |
| TBT | Technical Barriers to Trade |

TRIMs   Trade-Related Investment Measures
TRIPS   Trade-Related Intellectual Property Rights
UR      Uruguay Round
WTO     World Trade Organization

# Introduction

The threat of renewed protectionism is a permanent menace to the international trading system. While we have made tremendous advances over the last fifty years in opening markets for goods, services and foreign investment, the dangers of 'backsliding' towards more protection of our own markets is always present. There are still many countries, some of which are important trading nations in world markets, that have not yet even agreed to accept the international disciplines that govern global trade practices. Even countries that have accepted these disciplines have frequently used various legal provisions and 'loopholes' to contain foreign competition through additional restrictive measures against imports.[1]

The establishment of the World Trade Organization (WTO) and its dispute settlement mechanism has been a major step towards containing these protectionist dangers. The WTO Member countries have made remarkable progress in designing a system of 'capping' their space of policy changes for an *ad hoc* increase in protection. This has resulted in the establishment of far more stable trading conditions than those prevailing in the past. The Member countries have also agreed to enforce their commitment through legal adjudication and, if need be, through sanctions. However, as the events surrounding the WTO Ministerial Conferences in Seattle and Cancun plainly demonstrate, there are ample reasons to believe that the foundations of the international trading systems are not as solid as many would hope. The WTO system contains a number of legal provisions that many observers consider as instruments of new protection, such as the use of anti-dumping measures and safeguards and remedies. Moreover, as history teaches us, even a period of free trade should not lead one into complacency; in the last 150 years, the world economy experienced two periods of free trade, each being followed by relapse into protectionism.[2]

Given the dangers of protectionism, it is only natural to ask whether there are no other ways – in and outside the WTO – that would help minimize these threats and dangers. Is the current system of WTO disciplines sufficient to contain the dangers? If not, what changes would be required? Alternatively, can regional integration arrangements (RIAs) be conducive to the optimal conduct of trade policy? Can RIAs be effective in enforcing commitments made in regional agreements? Is there any difference in which RIAs operate? For example, does it matter whether RIAs are as sophisticated as possible or would it be preferable to have RIAs that are simple and straightforward?

These questions are frequently asked by academic economists, policy makers, international negotiators, journalists and all observers of the multilateral trading system. For many, the answer is 'no' to regional agreements. Regional activities have been subject to considerable controversies both in academia and among policy makers. Following the collapse of discussions in Cancun and previously in Seattle, as well as the proliferation of bilateral and other regional agreements, these questions are clearly very topical. Proponents of the multilateral trading system see regionalism as a threat to the multilateral trading system. Proponents of regionalism see these discriminatory practices as 'building blocks'[3] of global integration. Most economists strongly prefer unrestricted free trade but many also recognize that regional arrangements may be the second-best solutions. The debate has not been concluded and will undoubtedly continue in the future.

This book enters the debate from a rather different angle. While fully recognizing the merits of free trade, the book starts from the position that regional agreements can be justified as the second-best policy options in the presence of distortions. This simply reflects the current reality which is characterized by the proliferation of regional agreements around the world. Now, recognizing that regional models may differ, would it make any difference to the debate if we were to allow for differences in those models? The main theme of the book is that it does, indeed, make a difference whether countries conclude one type of regional arrangement as opposed to another. Deep integrations, in contrast to shallow integrations, are likely to be more successful in creating conditions for more efficient trade.

The main purpose of the book is to provide evidence about the effects of deep integration on policy making. We shall be arguing that deep RIA of a particular kind is conducive to good trade policy. By 'good' trade policy, we shall mean a credible and stable trade policy that avoids 'policy backsliding'. As noted, the latter has been a threat to both the international trading system as well as to financial markets and to flows of foreign investment.

In addition, this book examines *the reasons* why some RIAs are better than others in enforcing good trade policy. We shall compare three models of economic agreements – a simple ('shallow') RIA, the WTO agreements and a 'deep' RIA which we shall call 'WTO-Plus'. It will be argued that some RIAs – those that go deeper in coordinating government policies and complement more liberal trade policies – can be very effective in dealing with pressures for increased protection, and probably even more effective under certain conditions than is the case of the WTO today. The approach of the analysis is derived from economics rather than from politics. However, the reader may wish to complement the readings of this book by consulting articles such as Bussiere and Mulder (1999) who basically look at the same subject but from a perspective of political scientists. As a case study of a deep RIA, we shall consider agreements that have been signed by the Central European countries acceding to the European Union.

The book provides a fairly detailed review of trade policies of the Czech Republic, Hungary and Poland. These have been written by highly experienced trade economists from the region who have been in the frontline of their respective countries' policy making – M. Hrnčíř (Czech National Bank), J. Michalek (Warsaw University) and S. Meisel (Hungarian Academy of Sciences). The empirical evidence provided by our small sample of countries is supplemented with a review of the available evidence from other RIAs from outside the region, which is provided in Chapter 2, written by Cernat and Laird. We shall show that despite some 'slippages' in the conduct of trade policies, these Central European countries have done a remarkable job of maintaining the stability of their trade policies. The stability of trade policy is impressive, particularly in view of major disruptions to their traditional markets in the early 1990s and in view of the radical market-oriented reforms, which have lead to serious adjustment difficulties. Clearly some slippages occurred but these became far less frequent in more recent years than in the first half the 1990's as the countries progressively introduced measures more and more integrated into EU policy and institutional systems.

Chapter 1, by Drabek, discusses the theoretical differences between various models of RIA. Chapter 2 is an empirical presentation by Cernat and Laird. Chapter 3, by Begg (Birbeck College, London), looks at the role of fiscal policy under specific conditions of economic transition, and Chapter 4, by Erbenova and Holub (Czech National Bank), examines the role of macroeconomic policies in balance of payments management and as an instrument of trade promotion. The book concludes with the three Central European studies mentioned above.

## Scope: elements of 'good' trade policy

How do we define 'good' trade policies? This is clearly a critical question that needs to be answered before one can tackle the substantive problems. To do so, we shall start from the opposite end – that is, from what is not be treated in the discussion. The concept of 'good' trade policy can cover a variety of aspects. (1) For some, 'good' trade policy refers to the *level of protection* – 'low' protection for liberals, 'high' protection for 'protectionists'. (2) For others, 'good' trade policy means the *sequence* with which trade liberalization is introduced. For example, should trade policy liberalization precede other domestic policy measures, or at the same time, or be implemented only later? (3) The third interpretation of 'good' trade policy has to do with the choice of *instruments*. For example, should countries maintain quantitative restrictions, especially in the presence of market imperfections? Under what circumstances should countries adopt first-best or, alternatively, second-best policies? (4) Finally, 'good' trade policy can also mean its stability, which reduces the costs of uncertainty to investors as well as the costs of commercial transactions.

As already noted above, it is this last aspect – stability of trade policy – that is addressed in the book. By concentrating on the study of 'policy backsliding', our objective is rather narrow and clearly leaves out what many may consider other important aspects of 'good' policy.

In order to take the discussion further, another related issue that needs to be clarified is the precise meaning of trade policy 'backsliding'. If one is interested in 'backsliding', the next natural question of any observer is – 'Backsliding' in terms of what and from what?' What we mean in this study is 'backsliding' from countries' *international commitments* – be they in the form of commitments to other WTO Members or their regional trading partners.

Finally, there is the question of coverage. The empirical chapters in the book are concentrated on *border* measures. Non-border measures are also included but their treatment is somewhat rudimentary and a discussion of these measures in greater detail would make our exercise too technical and too comprehensive for the scope of this study. Nevertheless, we believe the loss of detail does not result in any significant loss to the substance of the argument.

## Enforcement of trade policy

'Good' trade policy can be enforced either unilaterally or through international cooperation. In most countries, domestic trade policy measures can be, and frequently are, enforced by domestic laws. Whether these legal safeguards provide sufficient conditions for stable trade policies remains an empirical issue. However, it is clear from the spread of RIA and from the creation of GATT/WTO that national legislations are not seen by the international community to be sufficient and adequate safeguards. For these reasons alone, trade commitments are more effectively enforced through international agreements.[4]

Unfortunately, the dangers of protectionism are not completely eliminated, even through most of the existing regional or multilateral agreements as discussed by Drabek in the volume. Both regional and WTO agreements maintain a variety of instruments which 'legalize' protection, which Hoekman and Leidy (1993) fittingly call 'holes and loopholes in integration agreements'. As Cernat and Laird show in this volume, these protectionist policy instruments have been used both at the regional and multilateral level.

These and related issues are discussed in Chapter 1 by Drabek. He starts with a summary of the main arguments for free trade and the reasons why some economists can live with regionalism as the second-best choice. Following a brief description of two broad regional models – deep and shallow – the author explains why certain RIA models have been effective in the conduct of trade policy as an instrument of enforcing government international commitments. The reasons are elaborated in more detail in the final section of the chapter.

## What the empirical findings show

The bulk of the contributions is focused on testing the hypothesis that certain 'deep' RIAs may be more effective in restraining countries from trade policy backsliding. We shall consider three accession countries and their Europe Agreements. As noted in the introduction, the countries include the Czech Republic, Hungary and Poland. The evidence is derived from original studies of Hrnčíř, Meisel and Michalek which were specifically prepared for the study on which this book is based. In addition, the chapter by Cernat and Laird provide relevant evidence from other RIAs that covers other regions.

### Stable level of applied protection

Following the dramatic reductions of trade protection in the region in the early 1990s, the level of applied protection remained stable over the period with perhaps only a minor trend towards downward adjustments. One explanation is that the *initial level* of protection was relatively high in two of the three countries under study – Hungary and Poland. In addition, Poland's tariff was initially unbound (Michalek), and the country only bound its tariffs in 1994 as a result of the Uruguay Round negotiations. Hungary did not eliminate the remaining quantitative restrictions until 2001; also its global quota on consumer goods was not dropped until the same year (Meisel 2001). The Czech Republic started its trade liberalization in 1991 with global quotas on agricultural imports, but those were eliminated by the end of the same year.

The stability of applied tariffs can be, to some extent, explained by the extensive level of bindings of these countries in the WTO. Most *applied tariffs* of these countries were actually bound in the WTO; for example, HG 95.7 per cent of total tariffs lines. Nevertheless, the countries retained some room to pursue more active trade policy. This was partly because some applied tariffs remained lower than the level of bindings. In Hungary, for example, more than 500 actual tariff rates were lower than the corresponding bindings (Meisel). In addition, some tariffs were not bound at all, as in the case of agriculture in Hungary, and this allowed these countries to use agricultural tariffs as a bargaining chip in later negotiations for accession. The only noteworthy exceptions to this stable pattern of applied tariffs were minor adjustments made at the time of the 'normalization' of relations in GATT (e.g. Hungary increased bindings on passenger cars, TV sets and some chemicals in 1990–91 and introduced the harmonized system (HS) and the EU combined nomenclature).

Another indication of fairly responsible and disciplined trade policy was the actual liberalization process which was started, and in some countries accelerated, after 1990. The most remarkable features of the process was the radical nature of the steps taken by governments and the *timetable* of the reforms. The timetable was very tight as the countries decided to pursue faster rather than slower reform. Moreover, the timetable was maintained in each country with, once again, relatively minor exceptions (Meisel). However, in Poland the process took longer and the government did not stabilize its external tariffs policy until the end of the Uruguay Round. Up to that point, tariffs were raised several times. This was possible because the tariffs initially were not bound in the GATT (Michalek)

**Limited use of non-tariff measures**

The relative stability of applied tariffs was, of course, helped by the countries' membership and their commitments in the WTO. They have eliminated all quotas and other quantitative restrictions, which represented the backbone of trade policy under central planning, and in line with the relevant GATT/WTO disciplines. Moreover, they have made only a very limited use of other non-tariff measures (NTM) including those that are WTO-legal, as the following brief review shows.

*AD and safeguards*

These measures have been in place in Hungary but not used, except against non-EU members. Poland applied safeguard clauses against imports from EFTA and the EU in the telecommunication sector and imposed special customs duties under the so-called Restructuring Clause of European Agreement in 1994. The other two sectoral safeguard clauses were applied in 1996 and again in 1997. Poland also drew extensively on special safeguard measures related to agriculture and agreed under European Agreement.[5]

*Quality, labelling and safety requirements*

In the area of consumer protection laws and environmental standards, the room for manoeuvre to use these instruments as a protectionist device has been shrinking as the rules are increasingly harmonized with the European Union.

*SPS and veterinary regulations*

SPS and veterinary regulations were used several times by Hungary against the European Union exports as well as against non-EU countries. Once again, these measures were be fully harmonized with those of the European Union by the time of the accession. On the other hand, Hungary offered a smaller cut in export subsidies in agriculture.

*Export licensing*

This was required in Hungary in the early 1990s to prevent domestic shortages (entirely in agriculture) but completely eliminated thereafter.

*Import bans*

These are very rare, although Hungary issued a ban on imports of cars older than four years (1995–2000). Similar policy was applied in the Czech Republic to help Volkswagen in the early stages of the company's investment. There were also product-specific import bans in the Czech Republic in the early 1990s (agriculture).

*Import surcharges*

For balance of payments reasons, Hungary applied import surcharges in March 1995, the Czech Republic in 1991 and Poland in 1992 (lasting until January 1997).

*Anti-dumping*

The only case of an anti-dumping measure reported by the authors was in the case of the Czech Republic on imports of salt in 2001.

*Other instruments*

Foreign currency deposits for imports of certain consumer goods were briefly introduced in the Czech Republic for balance of payments reasons. They were soon contested by the country's Czech trading partners and quickly abandoned.

*Other country experiences*

The sample of countries is quite small which makes it difficult to generalize. It is for this reason that the volume includes the chapter by Cernat and Laird, which provides additional evidence from other countries. As noted above, they have looked at various RIAs among developed and developing countries to see whether any particular model of RIA has been more effective. Their classification of RIA is somewhat different than the one used in the assessment of the Europe Agreements of the Central European countries. Nevertheless, their conclusions are broadly in line with the conclusions reached for the Central European countries – deeper integrations have a better chance of enforcing countries commitments in trade policy.

It would have been particularly interesting to extend the analysis to countries such as Bulgaria and Romania who also signed cooperation agreements with the European Union, known as Association Agreements, but those agreements sometimes may not have been as far-reaching as those of the countries in our sample, but they were fairly similar. Their inclusion in the book would be interesting but also expensive for this study.[6] It is interesting to note that even in the case of these two countries, the closer integration with the EU has resulted in a considerable trade liberalization and increased macroeconomic discipline. The latter was translated in Bulgaria into the adoption of a currency board with close links to the EU through capital inflows, trade and the euro as a reserve currency.

From the fragmentary evidence available in the literature, it is possible to note that in contrast to the three Associate countries in our sample, transition countries outside the accession 'orbit' appear to have experienced much greater changes in trade policy. For example, several countries in the Commonwealth of Independent States (CIS) – none of which had the kind of agreement with the EU – re-introduced fairly dramatic trade controls. For example, Russia re-introduced tariffs on exports of oil, gas, petrochemicals, metals, and various other goods in the course of the 1990s on several occasions. Kazakhstan introduced an import ban on some Russian products and set extremely high tariffs of 200 per cent on imports of goods from Uzbekistan and Kyrgyzstan. Uzbekistan quickly retaliated to these measures.[7]

**Critical role of macroeconomic policies**

How can we explain the remarkable stability of trade policy despite dramatic shocks – both external and internal – that the countries experienced and had to cope with? The authors provide a number of explanations but the stability can be mainly explained by two factors – 'good' macroeconomic policy and an effective enforcement mechanism with regard to trade policy commitments.

*Towards harmonization of macroeconomic policies*

The conduct of macroeconomic policy was a very sensible measure – at least in terms of the objective of maintaining external (balance of payments) equilibrium. As shown by Drabek in his review of the literature, one of the most critical elements in the package of polices that are subject to coordination by regional partners is macroeconomic policy. These are discussed in the book in two contributions. The limitations of traditional macroeconomic policies as instruments of balance of payments adjustment are discussed by David Begg. He shows how trade policies can be used as an alternative tool of balance of payments management under very restrictive conditions of severe structural maladjustment. In his model, tariffs play an important role of adjustment and prevent exchange rate overshooting. At the same time, however, fiscal policy remains a powerful instrument of balance of payments adjustment on the condition that extra resources are channelled into investment. Countries that do not have an adequate capacity to conduct an effective fiscal policy are in trouble on all fronts, including in the management of balance of payments.

An assessment of macroeconomic policies as a complement of trade liberalization is provided in this book by Erbenova and Holub. They look at the case of one of the countries covered by the authors of the empirical papers – the Czech Republic. With the help of an econometric model, they analyse the role of monetary and exchange rate policies as instruments of maintaining internal and external balances. They identify serious problems in the conduct of monetary policies and fiscal policies that led to the financial crisis in the second half of the 1990s. These policy distortions were subsequently corrected, and as the authors powerfully argue, this allowed the government to avoid taking protectionist measures.

*First-best policies of adjustment*

The policy makers in the region also turned out to be sensitive to the origins of balance of payments crises, which allowed them to discriminate between the *first-best* and second-best policies. There was a sufficiently early recognition among policy makers that certain balance of payments crises were actually of financial origin as opposed to 'real' crises.[8] For example, the Czech authorities in 1996/7 chose to restore external equilibrium through restrictions on capital movements *rather than* through the imposition of trade restrictive measures. This comes out very clearly from the analysis of the crisis period in the Czech republic by Hrnčíř.

*Flexible exchange rate policies*

The monetary authorities in all three countries understood well the merits of more flexible (typically, adjustable peg) exchange rate policies. The exception was the Czech experience, where the authorities insisted on a fixed nominal exchange rate for too long (Holub and Erbenova). The exchange rate policy was essential in the early 1990s due to the fact that the authorities were both unable and unwilling to enforce a strict macroeconomic coordination with the European Union. Despite pronouncements by the Hungarian authorities, the exchange rate policy served as anti-inflation instrument; competitiveness and structural adjustment were more important (Meisel). The policy was changed later, reflecting different priorities (i.e. anti-inflation). Similarly, the Czech government dropped its widely criticized policy of fixed peg of 1993–97 and moved to wider bands (Hrnčíř).

*'Entry' level of exchange rate*

Interestingly enough, the authorities were also very sensitive to the importance of the 'entry' level of exchange rate (Hrnčíř). Clearly, it makes a big difference if a country joins a monetary regime with an exchange rate which is perceived to be 'undervalued' as opposed to being 'overvalued'. The Czech experience is, again, a case in point. The Czech authorities fixed the original exchange rate at a level that was seen to make the koruna highly 'undervalued'. The 'undervaluation' (*polstar*, in Czech) giving a 'blanket protection' to domestic industries for adjustment and stimulating their restructuring and re-orientation of trade – clearly, an alternative policy to that proposed by Begg noted above.[9] In addition, 'undervaluation' was justified on the grounds of low foreign exchange reserves and the likely real exchange rate appreciation that was expected as a result of the growth of productivity (the 'Ballassa–Samuelson effect'). A similar pattern can be observed in varying degrees in all countries in transition.[10]

*Correction of macro-economic policy slippages*

The authorities in the three countries have taken commendable steps to correct mistakes in the conduct of macroeconomic policies. For example, the poor conduct of fiscal policies in the Czech republic was rectified in the mid-1990s through various fiscal policy packages, and there is a similar pressure to do so at present time.

*Importance of tariff revenues*

The conduct of macroeconomic policy would have been far less successful if the countries' budgets were excessively dependent on tariff revenues. However, this was not the case. Tariffs played a relatively small role as a source of budgetary revenues – except in Poland in the early 1990s. Nevertheless, it was only in 1991 when the government increased tariffs that the increase was motivated by budgetary difficulties (Michalek).

**Other instruments of making trade policy work better**

*Institutional reforms*

Rather than embarking on the road of protection, all three countries pushed actively towards institutional reforms as a way of trade and export promotion. It is well known that all transition countries suffered from serious institutional deficiencies of central planning. Institutional barriers such as problems in the banking sector – especially in handling international transactions and at reasonable costs, the lack of effective export credit guarantee schemes and other forms of export promotion had been the biggest constrains. This come out very clearly in other specialized studies such as MacBean (2000). Customs delays, bribes, poor physical infrastructure further added to the problems of the trade sector (Beilock 2000).

All of these issues have been addressed through wide ranging reforms. They are discussed in detail in this book by Erbenova and Holub in the context of their treatment of export promotion polices in the Czech Republic. It is quite clear from their contribution that the Czech government did not understand the institutional problems at hand but also made the necessary steps to ensure that the institutional constraints were mitigated.

*Domestic and foreign lobbies*

Lobbies and special interests are always a major factor in determining trade policy, and they have also been present in Central Europe. However, this is another surprising conclusion of the authors, these lobbies played a somewhat less important role than one often witnesses in other parts of the world. Protectionist lobbies appear to be rather more powerful in Poland and probably less so in the Czech Republic and Hungary. This itself would, of course, explain a large part of the trade policy behaviour of the three countries. Foreign lobbies, especially those related to foreign investors, also played a role. For example, Poland increased its external import duty on passenger cars at the time as signing the Europe Agreement (1992) as a part of the government policies to attract foreign investment into the car manufacturing sector (Michalek). The Czech government provided temporary protection to Volkswagen when the company acquired the Czech manufacturer Skoda. Similarly, foreign lobbies played an important role in the protection of the steel industry in Hungary (Meisel).

*The effectiveness of the enforcement mechanism*

The Europe agreements did not provide for a powerful dispute settlement mechanism along the lines of the system operating in the WTO. The agreements only called for consultations and the amicable resolution of disputes. Nevertheless, the countries, trade relations with the European Union have not been marred by many internal trade disputes related to a violation of trade disciplines. How do we explain this relative 'peace'? Similarly, the agreements did not provide for any coordination of macroeconomic policies, yet all three countries embarked on such a coordination more or less forcefully by the end of the 1990s. How do we explain this eagerness even though the coordination was not subject of the countries' international commitments?

The explanations are, of course, very simple – the anticipation of the countries' accession to the European Union acted as any most effective dispute settlement mechanism, if not better. In the anticipation of joining the European Union as soon as possible, the countries unilaterally decided to harmonize their monetary policies with those of the European Central Bank. They have pegged their currencies more forcefully to the euro through managed floating, and this has minimized, if not completely eliminated, problems that arise from maladjustment of currencies.

The 'peer' pressure emanating from the European Union also played a role in maintaining trade policy stable. When the Czech government decided to introduce import deposits in the face of its financial crisis in the second half of the 1990s, it was the European Commission that put pressure on the Czech government to remove the measure on the grounds that it discriminated against small companies. The measure was soon abolished and replaced with a conventional import surcharge.

## Conclusion

Judging from the empirical evidence provided in this book, the experience of the Central European countries appears to be quite remarkable and special. It was certainly exceptional in the light of the major disturbances and shocks these countries faced in the 1990s as a result of the collapse of central planning. The countries have maintained a fairly stable trade policy and demonstrated their eagerness not to renege on their international commitments. Trade policy slippages were only temporary, and perhaps only in Poland were they of some significance in the first half of the 1990s.

Why did the countries perform the way they did and so well? The evidence emerging from the papers points to two major factors: strong macroeconomic policy, especially monetary policy which 'shadowed' the policy of the European Central Bank, and a very effective enforcement mechanism that functioned on informal structures as well as peer pressure rather than a formal dispute settlement mechanism. Poland provides a 'classical' support for the argument. Arguably, the most significant trade policy slippages among the three countries were in Poland in the early 1990s, when the country experienced serious macroeconomic instability.

The second powerful factor was the enforcement mechanism of international commitments of these countries. The commitments vis-à-vis the European Union were part and parcel of these countries' new emerging political and economic links with the Union and their ambition to join the Union as full members. This also seems to have dictated these countries' behaviour as trading partners of the Union. Interestingly enough, trade policy slippages were more frequent at the time of the Europe Agreements. While the latter were wide ranging, they did not cover macroeconomic policies. What seems to have made a big difference was the moment when the countries began to implement the *Acquis Communautaire*, which is the primary driving force for the coordination of regulatory policies. Both the coordination of macroeconomic policies and the enforcement mechanism would tend to suggest that 'deep' integration is significant to maintain stable trade policy and to prevent trade policy backsliding.

The occasional slippages could be explained by a variety of factors. The level of initial protection mattered – it was very low in the Czech Republic and Hungary. Poland started with an initial protection that was relatively high but it was reduced over time. The reduction in border protection might have encouraged the government to rely from time to time on non-border measures. Premature trade liberalization was also blamed for increased protection of the steel industry in Hungary. The presence of a powerful foreign company has also been seen as a factor since powerful foreign companies are 'better placed for requesting and arguing for protective measures than traditional domestic firms' (Meisel).

**Policy implications**

What policy conclusions can be drawn from our study? First, one critical question faced by the Central European countries was at what stage should they harmonize their tariffs and trade policy instruments with those of the European Union. Should the harmonization be done before or after the accession? There is a divergence of opinion on this issue. Some argue that the policy coordination should take place before the accession – presumably as a demonstration of readiness to join the European Union (Meisel). In contrast, a relatively higher level of protection may serve as a 'cushion' at the time of integrating currencies – a goal of each of the acceding countries.

Second, what role should trade policy play at the time of balance of payments crises? All three Central European countries responded to their own crises by imposing import surcharges at one time or another. Protection of balance of payments was critical in all three countries. Was that the right policy? Should they have worried in the first place, for some experts believe that not all 'deficits are bad'? While the emphasis was initially put on import surcharges, the countries eventually shifted the policy towards tolerating current account deficits, provided these were generated by rapid growth of imports of investment goods.

Third, the success of the Central European countries in maintaining a stable *and* relatively liberal trade policy makes it tempting to suggest that the closer ties of these countries with the European Union have been highly conducive towards the success. There seems to be a clear evidence from these countries to support the argument that both the informal harmonization of monetary and exchange rate policies and the enforcement mechanism for honouring the countries' international commitments have provided the necessary conditions for government to avoid trade policy measures as instruments of balance of payments management and/or protection of domestic industry. It may be interesting to note that the WTO also operates with an efficient enforcement mechanism, but it is far less effective in dealing with protectionist measures imposed for balance of payments reasons. In other words, 'deeper' integration of the Central European countries seems to have worked well in integrating them not only into the European Union but also to the global economy.

This is, of course, far from saying that the policy of these countries to focus on their integration into the European Union is also the policy that has led to the *optimum optimorum* in terms of the countries' welfare. The approach in this book is focused only on one aspect of trade policy – its stability. Even though stability of trade policy should be a part of the country's welfare, it is rarely treated that way. We have not considered in this book other aspects which typically provide the basis for making judgements about 'free trade' and 'regionalism'. In particular, we have not considered at all the issue of gains from trade as typically defined in economic textbooks. Without a detailed knowledge of both static and dynamic gains that emerge from the integration of these countries into the European Union it is hardly possible to argue that 'deep' integration is more efficient than other forms of trading arrangements. However, when it comes to 'policy backsliding', the case for 'deep' integration appears much stronger, at least if it is based on the experience collected in this book.

## Notes

1   For evidence see, for example, Drabek and Brada (1998), which reviews the use of WTO – legal safeguards by countries in transition in the 1990s. Evidence on the growing incidence of anti-dumping measures can be found in Miranda *et al.* (1998).
2   See, for example, Boltho (1996).
3   The original question whether RIAs are 'building' or 'stumbling' blocks is attributed to Jagdish Bhagwati. For the history of the debate, see Bhagwati (1999).
4   There are, of course, other reasons why cooperative solutions among countries are preferable to unilateral measures as abundantly discussed, for example, in the literature on optimal tariff.
5   See the discussion on the critical role of macroeconomic policies in the next section. These measures *did not involve raising tariffs* but a decision to delay lowering of tariffs.
6   The EU–Romanian and EU–Bulgarian Agreements were *bona fide associations*. Rather than 'cooperation' agreements. Nevertheless, they contained similar elements of integration and trade liberalization. For instance, Romania has made a considerable progress in trade liberalization as part of its Europe Agreement as it eliminated most quotas and reduced tariffs. On the other hand, the agreements of the EU with CIS are very different.
7   See Hare (2001), p. 490.
8   The origins of the Czech crisis, for example, were attributed by most observers to excessive growth of wages and to capital movements, the latter being partially a function of the exchange rate management. As discussed in Drabek and Brada (1998), the link between trade and other policies is very strong. If fiscal policy or other policies are inappropriate, the determination to protect a chosen exchange rate or regime can result in enormous pressures to introduce protectionist policies.
9   The Czech 'experiment' of 1990/91 was, of course, more complex. The devaluation of the Czechoslovak currency was accompanied by the introduction of an import surcharge, a new licensing scheme and a global import quota on agricultural imports. These were maintained only temporarily. Nevertheless, the recourse to import restrictions in addition to currency adjustment goes a long way along the recommendation of Begg noted in this introduction.
10   The initial depreciation of currencies created a wedge between nominal exchange rates and those that would reflect PPP in all CIT. For more details see, for example, Hrnčíř in this volume. For Polish discussion, see Michalek in this volume.

## References

Bhagwati, Jagdish (1999): 'Regionalism and Multilateralism: An Overview', in Bhagwati, Krishna and Panagariya, *Trading Blocks: Alternative Approaches to Analyzing Preferential Trade Agreements*.
Bhagwati, Jagdish, Pravin Krishna, and Arvind Panagariya (eds) (1999) *Trading Blocks: Alternative Approaches to Analyzing Preferential Trade Agreements*, Cambridge, MA: MIT Press.

Beilock, Richard (2000) 'Will New Roads Help? Institutional Barriers to International Transport in Eastern Europe and the CIS'; University of Florida, Centre for International Business Education and Research (CIBER) Working Paper no. 00–14.

Boltho, Andrea (1996) 'Return of Free Trade', *International Affairs*, vol. 72, no. 2, 1996, pp. 247–59.

Bussiere, Mathieu, and Christian Mulder (1999) 'Political Instability and Economic Vulnerability' Washington, DC: International Monetary Fund, Policy Development and Review Department, Working Paper 99/46, April 1999.

Drabek, Zdenek, and Josef C. Brada (1998) 'Exchange Rate Regimes and the Stability of Trade Policy in Transition Economies', *Journal of Comparative Economics*, vol. 26, pp. 642–68.

Hare, Paul (2001) 'Trade Policy During the Transition: Lesson from the 1990s'; *World Economy*, vol. 24, no. 4, April, pp. 483–511.

Hoekman, Bernard, and Michael Leidy (1993): 'Holes and Loopholes in Integration Agreements: History and Prospects', in Anderson, Kym and Richard Blackhurst (eds), *Regional Integration and the Global Trading System*, New York and London: Harvester/Wheatsheaf.

MacBean, Alistair (2000) *Trade and Transition: Trade Promotion in Transition Economies*, London: Frank Cass.

Miranda, Jorge, Raul A. Torres, and Mario Ruiz (1998) 'The International Use of Anti-Dumping: 1987–1997', *Journal of World Trade*, vol. 32, no. 5, October, pp. 5–71.

# Part I
# An Alternative Overview of Regionalism

# 1
# Regionalism and Trade Discipline

*Zdenek Drabek*

*It may be possible to have a customs union – and indeed that more ambitious European animal – single market – without monetary integration, but it may also be true that in the absence of monetary integration the preservation of an area of free trade will always be under threat.* (Corden, 1994).

Regionalism is arguably one of the most controversial topics in the World Trade Organization (WTO), as well as among policy makers and academic economists. With its core principles deeply rooted in the idea of non-discrimination – both with regard to foreign firms (the 'most-favoured-nations' principle) and to competition between foreign and domestic firms (the 'national treatment' principle) – the WTO and its Agreements treats regionalism as an exception to its basics. For many policy makers, the multilateral road to trade disciplines and improvements in markets has been the preferred course of action. For others, the current trend towards the negotiation of bilateral and regional trading arrangements documents their concerns about the nature and pace of multilateral negotiations and their perception of concessions they are likely to obtain from their trading partners.

The academic community, too, is divided. Most international economists favour arrangements based on free trade and see regionalism as a step back from optimal solutions to the most efficient allocation of resources. Other economists have identified specific advantages of regional trading arrangements. These merits are primarily twofold. The first advantage is that regional integration arrangement (RIAs) may in some circumstances create dynamic conditions for *growth* of trade resulting from scale economies and improvements in (total factor) productivity. The second advantage of regional arrangements may emerge in the presence of market imperfections, which will justify corrections by government interventions. In this respect, regional arrangements can play a positive role in addressing these externalities.

However, it is clear from the continued controversies about the role of regionalism that neither the 'dynamic argument' nor the 'second-best argument' noted above have been able to resolve the matter unambiguously. Many opponents of regionalism fail to find any convincing empirical evidence to dismiss the claims of 'regionalists'. The latter, in turn, point to the continued presence of market imperfections, such as uncompetitive practices of firms, lack of information, negative externalities and other distortions.

Given the enormous literature that has been written about regionalism, the question may be asked 'What new issue can be addressed in the debate?' This chapter is an attempt to contribute to these debates by looking at regionalism from a different perspective than is traditionally adopted in the literature. Rather than asking which approach to international cooperation is 'better' or optimal, we shall try to assess the performance of particular types of RIA. We shall be asking the following questions:

- Why do various regional trading arrangements seem to perform better than others?
- Do some of the RIAs perform consistently better than others?
- If RIAs fail, what are the reasons?
- Is there any difference in the way in which different RIAs operate? For example, does it matter whether RIAs are as sophisticated as possible, or would it be preferable to have RIAs that are simple and straightforward?
- Given the strong preference of economists for 'free trade', can regional integration arrangements (RIAs) of a particular type be conducive to optimal conduct of trade policy in areas other than the two aspects mentioned above?

The aim of this chapter is to provide answers to these questions. We shall not discuss the questions in a broad sense. Instead, we shall limit the discussion to a very narrow issue of commercial policy, namely that of 'policy backsliding'. In other words, rather than revisiting the 'big question' of regionalism versus multilateralism, we shall only ask 'Can RIAs be effective in enforcing commitments made in regional agreements?'

My working hypothesis will be that the effectiveness of international agreements crucially depends on specific elements that form the backbones of these agreements. We already know that the so-called 'deep' RIAs can be more effective in terms of removing remaining trade-distorting measures. But I shall argue that 'deep' regional trading arrangements may also be more effective in terms of protecting countries against their partners' 'back tracking' on their commitments. In other words, these ('deep') measures, under special conditions, may be more effective at imposing the trade discipline than simple RIAs.

The reasons for the effectiveness of 'deep' RIAs are: (1) a high level of coordination of macroeconomic policies; (2) a strong incentive to cooperate and to seek consensus due to an accompanying process of political integration; (3) a wide-ranging coordination of regulatory policies and measures; and (4) a wide ranging liberalization of trade flows, including the elimination of border measures and a reduction of trade restrictive inside-the-border measures. In the context of the European Union, which is very much targeted in this chapter and this volume, the drive for this coordination of policies is provided through the *Acquis Communautaire*.

These issues can be analysed in two different ways. One approach could be theoretical when one analyses the structure of particular agreements and assesses their likely effectiveness in the light of the relevant objectives that are set for these agreements. This approach will be attempted in this chapter. The second approach would be to carry out an empirical analysis of the costs and benefits of these agreements. By looking at the performance of different RIAs, one can try to assess their effectiveness in the light of the same objective. The empirical assessment is the subject of the separate papers published in this collection of essays, with results summarized in the introductory chapter to this volume.

Before addressing substantive points of the debate, it will be necessary to deal with several methodological issues. These include an explanation of the notion of 'deep' regional integration and of the approach that is adopted in this chapter. In addition, we shall delineate the scope of our discussion with regard to the meaning of 'good' trade policy and 'policy backsliding'.

The structure of this chapter is as follows. The first section briefly reviews the main elements of the economics of free trade. This provides a convenient introduction to the debate on the merits or pitfalls of regionalism. Next, we describe the different types of regional trading arrangements. A distinction is made between 'shallow' and 'deep' integrations. In the subsequence section we discuss the effectiveness of different regional arrangements. We then discuss the driving forces of regional arrangements, subsequently looking at the specific issue of how different regional arrangements can maintain the stability of trade policy. The chapter closes by providing a framework for assessing different regional models.

## Economics of free trade: basic elements

Why do economists like 'free trade'? For economists, free trade is seen to be the 'best' market arrangement for one simple reason – trade allows countries to gain from international exchange and free trade, together with other market characteristics maximize the gains. There are many different sources of gains from trade: (1) trade allows the expansion of production possibilities of countries; (2) trade allows countries to increase their welfare measured in terms of consumption; (3) trade allows countries to specialize and thus increase their efficiency and welfare through resource reallocation; (4) trade allows access to those resources that are not domestically available; (5) trade allows countries to learn from the rest of world (e.g. new technologies, production techniques, managerial know-how); (6) trade increases the exposure of domestic firms to international competition, which stimulates internal efficiencies of firms and makes them more competitive; (7) together with other market conditions, free trade leads to the most efficient allocation of resources in which specialization in trade and production is optimized and welfare is maximized in each country; (8) properties of world prices are highly desirable. Finally, (9) there are also strong political economic arguments in favour of free trade. Let us now briefly turn to each of these properties and summarize the main messages.[1]

### Expansion of production possibilities

International trade allows markets to expand. Domestic markets are by definition limited by the size of population, their incomes and tastes. When countries trade they have access to *more* consumers and *greater* purchasing power than they do when they have to rely on the domestic market. As output expands, firms will most likely be in a position to benefit from economies of scale, leading to decreasing average costs of production. The latter will, in turn, expand the amount of output firms will be able to produce with a given (fixed) amount of resources (production factors). To use the language of economists, the production possibility frontier will expand.

### Specialization

Trade will always provide incentives to specialization under any kind of market arrangement, irrespective of whether the arrangements are perfect or imperfect. This simply follows the forces of absolute and comparative advantages that encourage countries to specialize. Specialization will lead to reallocation of resources, expanding the countries' opportunities in production and trade and making them more competitive. In addition, consumers will have the opportunity to access those goods and services supplied from the most efficient suppliers.

### Consumption and welfare

If households' objective is to satisfy their consumption needs and households do so in such a way that they maximize their consumption and producers maximize profits, free trade provides the best conditions to meet that objective. Free trade provides incentives to consumers to obtain goods at lower prices and thus increase consumer surplus. The consumer surplus is reduced by tariff by the amount of producer surplus and tariff revenue.

### Consumer choice

Free trade will allow consumers and firms to access the range of products and services that the country does not produce, since it may not have the resources and technical capability to do so. No country in the world is in a position to produce a complete range of products or to supply the whole range of services that are available in the market. No country in the world has access to all the resources that are needed in the production of commodities or in the supply of services. The scarce commodities or resources embodied in commodities will have to be imported if the countries want to gain.

### Trade as an instrument of competition

Many markets are too small to allow a large number of producers to operate in the market at competitive levels. A lack of competition may lead, in turn, to uncompetitive practices by firms and to internal inefficiencies. Increased competition in the domestic market generated by imports and through competitive pressures on exporters in foreign markets will substitute for the lack of competition from domestic firms.

### Free trade as an optimal solution to resource allocation

Under certain conditions, free trade will be the optimal solution to resource allocation. Welfare of each country will be maximized, permitted by the most efficient allocation of resources. Trade specialization will be optimal under free trade. Market restrictions of any form will lead to (at least temporarily) trade diversion from more efficient suppliers to less efficient ones. In brief, free trade allows countries to maximize their gains from trade from specialisation.

### Learning by doing

Perhaps one the most underestimated sources of gains from trade is gains from learning by doing. Firms learn new managerial techniques, they learn about production methods and techniques, ways to access markets, distribution channels. The access to all of these properties is seriously impeded under autarky and protectionism.

### Properties of world prices

Prices formed in perfectly competitive world markets have a number of highly desirable properties. First, world prices are equilibrium prices. Being formed in free markets by forces of supply and demand and without interference of government measures, prices settle at a level at which demand is fully satisfied by the existing supply. While this property may appear too elementary, there are many examples in which the property has been violated. For example, the present transition countries conducted their trade in an institutional set up in which prices did automatically adjust to relative scarcities.

Second, the prices encourage firms and countries to increase output up to the point at which prices are equal to marginal production costs. When prices are equal to marginal costs, firms do not earn any extra (monopoly) profits and produce at the optimum rate of output.

Third, with perfect foresight and information, externalities are properly discounted in the pricing behaviour of firms and households. As a result, private costs are the same as social costs and private benefits are equal to social benefits.

Fourth, free market prices formed under perfect competition reflect true relative scarcities of goods and services. They are not affected either by government interventions such as tariffs or other trade policy instruments or by uncompetitive behaviour of firms.

### Political economy of free trade

Protection and other government interventions have important political implications for international policy making. Protection is always associated with lobbies that seek their special interests at the expense of other groups. Specific lobbies tend to increase domestic protection and make it more difficult for governments to remove trade barriers in the future. Protection and the presence of lobbies make domestic policy making less transparent and predictable. Non-discriminatory free trade eliminates all forms of discrimination reducing, therefore, the scope for pressures from different lobbies.

## 'Deep' regional integration: what does it mean?

In real life, free and perfect markets are more an exception than a rule. The presence of distortions means, of course, that there is room for government interventions. For example, if governments face world prices that are distorted by protectionist policies in their partner countries, the theory of 'second-best' allows for measures to be taken by governments to offset those distortions. Measures to establish regional trading arrangements can be used as such a second-best alternative. This is based on the simple idea that the effect of foreign restrictions can be more than offset by gains generated from more open trade regimes and markets in which only a limited number of countries participate.

There is a variety of different concepts and models of RIA that has been proposed in the literature. The differences reflect different criteria applied in the analysis of RIAs and the personal preferences of analysts. Woolcock (2003) has recently tried to provide a framework for assessing the impact of legal arrangements on the structure of RIAs. Their structures can differ in terms of (1) their coverage; (2) procedural measures; (3) substantive provisions; and (4) implementation and enforcement. Clearly, the criteria reflect personal preferences but they are helpful in more powerfully making the point that the *type/model* of RIA matters.

Broadly speaking, it is possible to identify at least two broad approaches.

### Free Trade Arrangements (FTAs)

The most common approach to RIA is free trade arrangements. There has recently been a proliferation of FTAs around the world, reflecting to a large extent attempts of countries significantly to expand the product coverage subject to complete or partial tariff reductions. For example, of all the RIAs notified in the WTO, 43 had 100 per cent coverage in industrial products in 1998, which compares to only 11 in 1990.[2]

FTAs can distinguish between two additional groups – *'geography-based RIAs'* and FTAs, which are not based on 'geography'. According to the proponents of the geography-based approach, it is geographical distance that matters in forming 'natural' trading partners. Countries are more likely to form RTAs with their neighbours than with distant countries. For example, Frankel and Romer (1996) argue that geographic characteristics of countries are critical factors in determining their trade and may be even uncorrelated with their incomes. The proximity to markets and the likely similarities in the way neighbours organize their societies is conducive to more trade and other economic relations, and it is also more conducive to political and economic cooperation. For these reasons, geography-based RTAs are often called 'natural' RTAs. Examples of 'natural' RTAs are ASEAN in South East Asia, the European Union, NAFTA, MERCOSUR and many others. In fact, most observers have probably thought of RIAs as those based on geography![3]

A semantic variety of geography-based models are models known as North-North, South-South and Mixed. Despite their names, these models of RIA do not actually refer to geography but, rather, to stages of development. Thus, North-North RIAs refer to those RIAs that are formed by developed countries. South-South RIAs are those that are formed by developing countries. Finally, North-South RIAs are those that are formed by developed and developing countries. The latter model has been alternatively called 'mixed'. The approach has been used, for example, by Cernat and Laird in this volume.

What makes these RIAs different from other RIAs is the range of policy instruments that is coordinated by members of the RIA in question. For this reason, we call these RIAs 'shallow'. Typically, the 'shallow' RIAs start and finish with the coordination of border measures. This may involve the elimination of all tariffs and import quotas in mutual trade of regional partners, leading to the establishment of a 'free trade area'. If the members decide further to coordinate external trade policies, that is policies towards non-members, this will lead to the establishment of a 'customs union'. For that reason, the latter may already be see as a form of 'deeper' integration.

### 'Deep' integration

The second broad type of RIA divides RIAs according to the depth of integration of the RIA's members. Thus, a distinction can be made between 'deep' RIAs and 'shallow' RTAs. Most recent RIA initiatives have been directed towards deeper arrangements; that is, going beyond the simple tariff-cutting exercises. But that is where the agreements among observers would end. For some, 'deep' RIAs are defined in terms of the coverage of economic sectors and activities. For example, following Lawrence (1996), Hoekman and

Konan (1999) define 'deep' integration as 'government actions to reduce the market segmenting effects of domestic regulatory policies through coordination and cooperation'.[4] The aim of the cooperation and coordination of members in such a case is making their markets more contestable. But the depth of integration could clearly also cover other areas. Arguably, the most interesting and important area of coordination is macroeconomic policy, for it can be shown that the stability of trade policy is critically linked to macroeconomic stability.[5]

There are different forms and shapes in which member countries can coordinate their macroeconomic policies. The simplest procedure would be a situation when a particular country chooses to coordinate its macroeconomic policies without any formal agreement. The tight links of the Austrian monetary policy to German monetary policy before the formal accession of Austria to the European Union is a case in point. The current close links between Swiss monetary policies and the policies of the European Central Bank provide another example.

More formal arrangements of a close coordination of macroeconomic policies are *currency* union and *monetary union*. When countries create currency union they agree to retain their currencies but with exchange rates pegged at a fixed level. This implies that governments fully subordinate other domestic objectives to the commitment of maintaining fixed exchange rates. An even deeper form of macroeconomic coordination is the formation of monetary union when countries agree to surrender their national currencies and to adopt single currency for the whole region.

In brief, 'deep' integration arrangements refer to measures that go beyond border measures. Which measures are adopted is subject to choice. For example, certain bilateral free trade agreements of the United States include labour or environment provisions in addition to provisions to eliminate tariffs and other border restrictions. The European Union member countries have chosen to coordinate domestic regulatory policies as well as monetary policies. Thus, the EU model actually assumes a structure based on the elements of monetary union.

## Effectiveness of 'deep' regional arrangements: a theoretical treatment

'How effective are regional trading arrangements?' 'What constitutes the main elements of successful regional arrangements?' These questions have frequently been asked by policy makers and in the academic literature. The conclusions of the debate have been somewhat ambiguous. Nevertheless, several arguments have been put forward to suggest that under specific conditions – often related to conditions characteristic for 'deep' integration arrangements – RIAs can improve national welfare. The critical assumption is, of course, that RIAs are only formed as the 'second-best' policy options, implying that their formation is intended to address a particular distortion of world markets.[6]

The benefits of 'deeper' integration have been recognized for a long time. In discussing the merits of border and inside-the-border measures in the European Union, Cecchini *et al.* (1988) emphasized the need to eliminate technical barriers and/or harmonize technical standards in order to obtain the full benefits of a true single market through economies of scale. His point was simple; a reduction in tariff barriers may do very little if firms cannot compete because of restrictions imposed on the technical specifications of their products. The argument of Hoekman and Konan (1999) to move beyond border measures towards harmonization of domestic rules is in the same spirit.

However, no one has probably argued more forcefully in favour of deep integration than Lawrence 1996. In the project *Integrating National Economies* of the Brookings Institution, he argued that one had to take a new and fresh look at regionalism, especially at the recent cases of free trade agreements. He makes an important point that modern trade theory is unable to assess the contribution of deeper integrations. The theory typically looks at welfare effects of border measures while deep integration also involves inside-the-border measures. The objectives of deep integrations may go beyond a removal of trade barriers; some integrations target the objective of increasing competition, others aim at increasing investment. He is careful to suggest that deep integration is *a priority*, neither better nor worse, but he is quite clear that deep integration can be more successful in targeting certain objectives.

The theoretical literature has addressed various aspects of regionalism and its effectiveness. The assessments depend, first of all, on the standard against which RIAs are evaluated.[7] The theory has recognized at least the following five standards. First, regionalism has been traditionally assessed in terms of static welfare gains. Following the original writing of Viner (1950), the issue is addressed in the literature by asking whether RIAs are 'trade creating' or 'trade diverting'.

Second, RIAs can be also assessed in terms of their effectiveness in *enforcing the common external policy* of its members (de Melo, Panagarya and Rodrik 1993). The proponents of this argument suggest that RIAs (customs union) are likely to dilute political powers of individual members, disrupt the formation of rent-seeking interest groups and stimulate more efficient outcomes by seeking compromises among members with a different set of priorities.[8] Some even argue that 'deeper regionalism gradually gives rise to *negotiated and rules-based political space* including confidence-building measures, common rules and procedures, political dialogue and co-operation'.[9]

Third, RIAs can be also seen as instruments of reducing *policy uncertainty*. The latter increases the risk premium for foreign investors and therefore, increase, the costs of borrowing as well as the opportunity costs of capital to investors. Fourth, RIA can be instrumental in creating an environment for better design and conduct of economic polices – *better governance*.

This point has been typically emphasized in the context of NAFTA and its impact on policy reform in Mexico. Fifth, regional arrangements can act as an instrument of trade liberalization. The question remains open, however, whether these arrangements, which are typically based on reciprocal agreements, provide more favourable *incentives* to trade liberalization than those required for unilateral trade liberalization or in the context of multilateral negotiations.

The fact that standards against which RIAs can be evaluated are different is critical. Clearly, if a group of countries decides to form a regional economic arrangement irrespective of costs, this may make the arrangement a success in political terms but it would hardly satisfy an economist who seeks to evaluate economic policies in terms of costs and benefits. The cost-benefit analysis is typically done in terms of the effects on countries' welfare, which is measured in terms of producer and consumer surpluses and government revenues.

The static economic argument about regionalism is primarily dependent on the balance between 'trade creating' and 'trade diverting' forces within the RIA.[10] The balance will be determined by a number of country-specific factors. These include, in particular, the initial level of protection, the existence or absence of rules of origin,[11] geographical distance and other geographical characteristics, power politics within the RIA, mutual shares of member countries' trade, trade policy towards non-member countries and the role of internal trade barriers. In brief, the impact of a particular RIA on economic welfare is an empirical issue – under certain circumstances, on balance RIA will be trade creating, while under other circumstances RIA will be trade-diverting.[12]

Additional complication will arise once we allow for different concepts of economic welfare. For example, should not country's economic welfare also include effects of policy uncertainty which may be due to unstable trade policy (and a government's proclivity towards 'policy backsliding') or a lack of transparency? It should be evident that policy uncertainty leads to higher risks and, usually, to higher costs of doing business. This qualification has an important bearing not only for the evaluation of RIAs in general, but also for the evaluation of trade policies of the acceding countries to the European Union in particular.

When we limit the analyisis to the convential assessment of welfare gains, the guidance of theory about different models of RIA remains limited. Most of the debate has been concentrated on the effectiveness of free trade areas (FTAs) and customs unions. FTAs have been traditionally viewed with scepticism as an instrument of trade creation, primarily because of the existence of rules of origin (ROO) which tend to increase costs of production (Krueger 1993). For example, a recent study of NAFTA by Anson *et al.* (2003) shows that up to 40 per cent of Mexico's preferential access to the US market in 2000 was absorbed by ROO-related administrative costs. Further evidence of high costs of ROO is provided in other studies such as Ganay and Cornejo (2002).

However, what the debate also shows is that the simple RIAs that only involve the removal of internal border restrictions are dysfunctional. FTAs will not typically survive without the additional element of policy coordination – the introduction of ROO. The presence of preferential ROO will not necessarily have a reducing effect on welfare. The critical condition for FTAs to be welfare enhancing is that goods produced within the union – whether final or intermediate – be allowed to trade freely and that the partner countries cooperate beyond a simple elimination of internal tariffs. The point is made in a paper by Panagariya and Dutta Gupta (2001), who also make a related point in support of free trade areas (FTAs) when they argue that rules of origins can improve the political viability of FTAs.[13]

Somewhat more explicit treatment of different regional models has appeared in the theoretical literature on optimal tariff. In theory, countries can impose an optimal tariff to compensate for losses due to distortions in world markets. Alternatively, large countries may be in the position to exploit their market power and improve their terms of trade by levying an optimal tariff. One of the important questions is whether the level of optimal tariff differs under conditions of a regional trading arrangement, as opposed to the situation in which such arrangements are absent. Another question is whether the welfare gains resulting from the levy will offset losses that may occur due to trade diversion after the formation of a regional arrangement. In addressing these issues, Schiff and Winters (1998), for example, constructed a model in which RIAs unambiguously raise welfare by correcting for externalities. They assume that the regional trading arrangement increases trust among countries participating in such an arrangement – generating positive externalities in the process – and reduces the likelihood of conflict. They show that the optimum external tariff on imports from the rest of the world will tend to decline over time and that deep integration implies a *lower* optimum external tariff in typical situations. Other papers have been reviewed by Winters (1999), who finds that the outcomes critically depend on the parameters and specifications of individual models.[14] In these cases, the attraction of RIAs stems from the extra coordination by member countries involving a co-operation to correct for externalities and, as in the case of the treatment of ROO in the model of Panagariya and Dutta Gupta, from political cooperation.

The most explicit treatment of regional models comes out of the theory of customs union. A relevant argument has been recently made by Panagariya (2002). Building on the Kemp–Vanek–Ohyama–Wan proposition, which states that if two or more countries form a customs union by freezing external tariffs and eliminating internal trade barriers, he shows that the union as a whole and the rest of the world cannot be worse off than before. Thus, by focusing on trade policies not only towards members but also on non-members, the theory of customs union extends the argument beyond the assessment of static gains from the removal of trade barriers among members. I call this a 'primitive deep' integration.

In short, the economic theory of RIA makes only a limited, and often only indirect, distinction among different types of RIA. The 'optimal tariff' literature, for example, includes more complex models that can be interpreted with reference to 'deep' integration. The extension of models to include different elasticity of substitution between domestic and imported goods, or to expand the coverage of activities, or the inclusion of models that assume coordination of trade policies by regional partners towards the rest of the world are examples of such initiatives.[15]

In contrast, the issue of 'deep' integration has received a far greater attention in the political theory and in the broader area of political economy, especially under the heading of fiscal federalism. It is not the intention to elaborate those theories in this chapter. It may only suffice to say that the concept of 'deep' integration has received a far more sympathetic and positive treatment than that accorded by pure economic theorists. The positive treatment primarily originates in the recognition of sizeable benefits that can be obtained from the provision of public goods. The latter can be of either economic or non-economic nature. The benefits come from increased scale or the scope of activities in the presence of spill-overs. Costs of 'deeper' integration come from information asymmetries and heterogeneity of the geographical unit.[16]

## Effectiveness of 'deep' regional arrangements: an empirical support

The positive effects of 'deep' integration have been identified more clearly in the empirical literature. If the regional arrangements also target financial markets in addition to trade policy measures, the objective is to create currency/monetary unions. The financial integration takes place when countries introduce common currencies and/or give up monetary independence, and as securities markets become deeper, more liquid and as transaction costs fall. Studies of the European Monetary Union (EMU), for example, show that the creation of the EMU has had a favourable impact on the integration of capital markets. For example, Danthine, Giavazzi and von Thadden (2000) or Portes and Rey (1998) argue that Euro assets have become far more efficient and attractive as a result of this process. Furthermore, Dee and Gali (2003) show that a simple preferential trade area is not enough – after controlling for a number of factors – to generate incentives strong enough to augment 'traditional' investment significantly in relation to trade flows.

Others have also argued that deep integration will lead to convergence of corporate practices. As (intra)regional economic links expand, firms become more concerned about the links between their legal origins and investment protection and finance.[17] The rapidly growing finance literature points in the direction of changes in corporate governance towards more harmonization. For example, with evidence provided from transition and developing countries, Bergloff and von Thadden (1999) identify forces that lead to the convergence in the pattern of corporate governance, particularly in countries with strong and regional integration commitments.

Significant economic benefits of deep regional integration are expected to come through the trade channel. There has been much debate about the impact of different regional trading arrangements on the level of trade which, in turn, stimulates the growth of domestic incomes and affects national welfare.[18] The trade effects together with the effects of RIAs on investment have been the subject of numerous studies. Perhaps most of the research on RIAs has been done on the European Union – arguably the best example of a 'deep' integration – and its effects on trade and investment. The majority of studies have shown that trade creation has been far greater than trade diversion in the case of manufacturers, but the pattern was reversed in the case of agriculture (e.g. WTO 1995). The trade creation effect was also most certainly prevalent in the case of the newly acceding countries from Central and Eastern Europe.[19]

Even more powerful arguments and evidence has come from the more recent empirical literature on trade effects from those 'deep' regional arrangements that involve coordination of macroeconomic policies. The literature is consistent and mutually supportive in the conclusion that coordination of macroeconomic policies together with trade liberalization are extremely conducive to growth of trade. For example, studies on the impact currency unions on trade (such Frankel and Rose (2000) and the more recent study of Parsley and Wei (2003)) strongly emphasize the powerful role of currency unions on growth of trade. The latter study goes even further when the authors argue with the support of empirical evidence that countries joining a currency union in which the member countries' currency exchange rates are fixed and supported by monetary authorities stimulate trade as much as or more than free trade arrangements.[20]

It could be argued that currency unions represent optimal currency areas, the efficiency of which is determined, *inter alia*, by distance. In other words, the strong correlation between growth of trade and the presence of common currency could, in theory, be explained by the proximity of the countries that establish currency union.[21] If one were to ascertain the role of geographical factors as a determinant of trade flows of currency unions, one should, therefore, control for distance in the estimating equations. This has been undertaken in a recent study by Greenaway and Milner (2002), who find positive trade effects of RIAs even after controlling for distance.

Deeper integration has made a big difference to the way in which countries specialize and firms invest. The unprecedented expansions of 'production fragmentation' divides the value chains of production processes into distinct stages. This has led to 'disintegration of production' into different geographical locations which are integrated by international trade. Such a fragmentation of production is only possible when countries emphasize product standards, innovation, adaptability and the speed of response. This, in turn, requires appropriate policy, and regulatory and other institutional support from the authorities. Everybody gains in the process; the large (foreign) investor gains by accessing cheaper sources of supply, and the small firms in most countries gain by accessing modern technology and larger markets. The important and relevant point is that regional integration can be highly conducive to the effectiveness of this process. For example, as shown by Kaminski and Ng (2001), most transition countries acceding to the European Union or in the process of accession have made huge progress in adjusting their production structures and infrastructures to these requirements, particularly as a result of legal and institutional harmonization with the European Union. The spectacular expansion of trade involving China, Japan and the other South East Asian countries has been a powerful stimulus for them to seek closer cooperation in the area of reserve management and exchange rate management, and discussions have even started about cooperation in liberalizing financial markets.[22]

What is common to all these arguments is the role of more comprehensive regional trading arrangements. The impact of RIAs will be more likely trade-creating, *ceteris paribus*, if the RIA in question is more comprehensive/deeper. Panagariya's explanation of gains from customs unions noted above rests on the existence of rules of origin. The argument offered by Schiff and Winter can be also interpreted as an argument in favour of more comprehensive RIAs that promote greater political harmonization and cohesion.

On the other hand, it must be recognized that the arguments in favour of deep integration can only be made subject to certain provisions. Lawrence (1996) himself, who is a strong supporter of 'deep integration' as noted above, recognized that 'deep integration' is a guarantee neither to increased efficiency gains nor to other forms of successes. To quote, 'I do not mean to imply that deeper is better. Indeed, deeper integration could be better or worse, depending on the nature of the policies and the countries to which they are applied.' Lawrence (1996, p. 8).

The harshest criticism of regionalism comes from Bhagwati. He argues that there is no unambiguous empirical evidence to suggest that RIAs are more effective (faster) in opening markets globally. He points out, for example, to the protracted experience of the European Union with a reform of the Common Agricultural Policy. He also suggests that deeper RIAs are not necessarily more efficient in producing better results. On the contrary, these arrangements will lead to agreements in which 'weaker states agree to conditions imposed by stronger states' (Bhagwati 1999, p. 24) In his view, RIAs involving hegemonic relations between partners lead to inefficient solutions for the less powerful members of the relevant RIA through the impositions of exceptions, specific rules, and the tilting of the weight of changes in the conduct trade policy towards the interests of the more powerful member.[23] These views are colloborated by Ann Krueger and her work on rules of origin (Krueger 1993). Her work shows that rules of origin may act as a highly inefficient and protectionist instrument. This would imply that the proliferation of deep RIAs will lead to a proliferation of highly inefficient rules of origin that, in turn, lead to restrictive trade practices and loss of welfare.[24]

In sum, the theoretical debate of merits of regional trading arrangements and multilateralism remains ambiguous. However, there are strong empirical reasons to believe that deeper rather shallow integration makes a difference to the growth of trade and, hence to welfare effects – under *ceteris paribus* conditions. However, the conditions can and do vary in the real world, and the discussion in the following section is an attempt to identify at least some important conditions that can play a role in specific and highly confined circumstances.

## Driving forces of effective regional arrangements

Given the diversity of integration models, it is not surprising that not all integrations have been successful or effective. In fact, the history of RIAs is rather an extensive list of failures with 'successes' being an exception rather than a rule.[25]

What makes some RIAs effective while other arrangements have failed? This question has been addressed in the economic literature quite extensively, and the answers can be divided into several groups. Before elaborating the issues, it should be noted that the following discussion only refers to economic rather than to political factors.

### RIAs versus unilateral liberalization (UL) versus multilateral liberalization

The magnitude of the effects of trade policy critically depend on 'initial conditions'. If trade barriers are initially high and/or domestic distortions severe, regional arrangements may only move countries to marginally improved outcomes. The external barriers in trade with ROW will ensure that domestic, inefficient industries continue to survive. It follows that RIAs are not necessarily more effective than UL.

It is for this reason that countries with high trade barriers would probably benefit more from UL than from free trade arrangements (FTAs).[26] This further implies that countries with low trade barriers are likely to benefit more from FTAs or non-discriminatory agreements than from additional UL due to reciprocity and, as a result, improved market access. It further follows that the deeper are the measures to remove trade barriers, the greater is the chance that RIAs will be effective.

Many RIAs may have failed for similar reasons because they have not been successful in meeting their principle objectives. For example, RIAs may be seen as an instrument of attracting FDI. However, a recent study by Sapsford, Griffith and Balasubramanyam (2002) shows that it is not the presence of RIAs that determines the magnitude and direction of bilateral FDI flows but rather the economic characteristics of both the investing and the host country. Clearly, these economic factors include primarily economic fundamentals rather than outcomes of economic policies.

Another twist to the argument, this time in relation to global rules and arrangements, was given by Hallet and Primo Braga (1994). Drawing on game theory, they argue that RIAs can work better than global rules as pre-commitment devices for internally cooperative policies because they create a denser network or interlinked policy targets. Policy 'deepening' by expanding the list of variables covered by the trade agreement will tend to make the bloc more cohesive.

## Choice of policy variable matters

Choice of policy variables is relatively straightforward when the objective is to open up markets. If countries wish to eliminate all trade barriers completely and integrate their manufacturing sectors into the global economy, they will not only have to reduce trade barriers in industry but they will also have to coordinate other regulatory policies that affect the activities and performance of the manufacturing sectors. By way of an example, full integration of industrial sectors in the European Union could not be achieved simply by removing all tariffs and quotas on intra-EU imports. The process had to be accompanied by harmonization of technical standards, labour policies, competition policies and so on. These are measures that have been already recognized in writings of Hoekman and Konan (1999), Lawrence (1996), and others. The outstanding but difficult question is which regulations are 'trade-promoting' and which are 'trade-distorting'.

The critics of regionalism point to the continued presence of selective and elastic instruments of protection in RIAs. This is what Hoekman and Leidy (1993) called the 'holes and loopholes' of international trade agreements. These measures include, for example, anti-dumping measures, voluntary exports restraints and, as already noted above, trade-restrictive rules of origin. These 'holes and loopholes' are not only the privy of RIAs but also of the WTO agreements. It is for these reasons that Bhagwati and others have suggested a revision and greater disciplines in the GATT Agreement to be imposed on anti-dumping as per Article VI of GATT and a greater compliance with Article XIX[27]. Bhagwati (1999) has also proposed an appropriate revision of Article XXIV that regulates regional arrangements to enhance clarity and ensure that RIAs are trade-creating.[28] Many of these and other related issues have been already taken up in the Committee on Regional Trade Agreements but so far without any conclusion.[29]

## Choice of partners

Choice of partners also matters. The debate about the choice of partners for international coordination has been focused primarily on the rather controversial issue of *natural* trading partners, for reasons which are further elaborated below, and on the size of RIA.[30] The issue about natural trading partners is whether natural trade relations contain features of trade and other economic relations that are inherently conducive to trade among specific countries. As already noted above, transport costs constitute an important element of natural trading links. Low transport costs will, *ceteris paribus*, stimulate trade and *vice versa*.

Suppose, for example, that a country has an opportunity to form a regional trading arrangement with two different blocs of countries. One bloc includes countries with a high level of income and technological attainment but who have not had intensive trade relations with the country in question. The other bloc includes countries with a low level of income, slow growth, poor technology but who represent a traditional trading partner of that country. In choosing between these two blocs, the authorities will have to consider the likely adjustment costs resulting from preferential trading with each of these blocs, the opportunity costs of choosing one bloc over another (i.e. foregone benefits from lost opportunities in the other bloc), the room and scope for further adjustments in the future to maintain competitiveness (i.e. the ability to minimize the effects of entrenched lobbies) and so on.[31] Clearly, the authorities will also have to consider the size of markets in partner countries as well as the existing ties with a given country. The latter were analysed in a specific model by Schiff (1996) who argues, *inter alia*, that a country joining a preferential trading arrangement is better off as a small country in a large bloc rather than a large country in a small bloc. Furthermore, the country's welfare will be higher after formation of the arrangement if imports from its partner country are smaller, irrespective of whether the partner country is small or large.[32]

In his recent paper on the distribution of gains and losses from RIAs, Venables (2003) uses a formal model to assess the importance of factor endowments in generating effective agreements. He argues that the outcomes depend on the comparative advantage of members, relative to each other and relative to the rest of the world. Countries with a comparative advantage between that of their partners and the rest of the world do better than countries with an 'extreme' comparative advantage.

Income levels of countries are also likely to affect the outcome of RIAs. For example, as de Melo *et al.* (1993) point out, integration among low income, developing countries is unlikely to benefit from major efficiency gains. This is primarily because gains from specialization between partners will be limited. In contrast, similarity in income levels will reinforce the gains from regional integration.[33]

In brief, the choice of countries in regional arrangements matters for a variety of reasons. The size of countries will affect the size of markets. The initial level of protection and mutual trade will affect the net contribution of trade creation and trade diversion to the national welfare of each country. A number of other factors will affect the degree of complementarity or competitiveness of countries joining a RIA. For example, differences in factor endowments will increase the former, while similarities will increase the latter. These differences will, in turn, determine the relative competitiveness of firms and thus the costs of adjustments required to make the RIA in question work. The ability to coordinate activities in the political area will enhance the ability of member countries to coordinate their economic policies and so on.[34]

Whether by chance or by the influence of economic forces, the allocation, accumulation and location effects are seen by many observers to be most powerful among natural trading partners. For example, Summers (1999, p. 563) argues that 'existing and any contemplated regional arrangements link nations that are already natural trading partners'. Krugman and Venables (1990) show that firms in an imperfectly competitive industry will, during a process of integration, be drawn towards 'central' areas of the region because having good access to consumers becomes more important to firms as the costs of market access decline.

A policy implication of all these arguments about 'natural' trading partners is that the selection of partners will also make a difference for the establishment of a deep RIA, which cannot be made in a vacuum. Certain conditions will have to be respected for deep integration to work. This is evident from the various experiences of countries.[35] The choice is sometimes a matter of politics and security or a credibility of economic policies. Other experience also teaches that regional integration is as much about 'trade creation or diversion' as about costs of adjustment.[36] Which conditions are necessary to make deep integration viable? Unfortunately, economic theory does not provide good guidance, and the debate about deep integration remains rather general. Political will is undoubtedly a crucial factor to establish a successful deep RIA since any regional arrangement implies a loss of certain degree of national sovereignty. The RIA in question must also be economically successful, which typically implies a positive contribution of trade to national welfare. Currency union *per se* does not provide an unambiguous explanation of trade creation. Beyond these general conditions, however, the identification of specific factors is difficult.

### Role of institutions

De Melo, Panagariya and Rodrik (1993) make the powerful point that strong institutions that are common for all members of RIAs also matter. This, in turn, would require a relative commonality of members' objectives, which translates into a relative homogeneity of members' interests. Clearly, as others have pointed out, the homogeneity has been the prime reason for the success of the European Union so far (Wyplosz 2003).

Historically, most RIA have been focused on free trade agreements. This has virtually left untouched the question of institutional cooperation and deepening. Their sole objective was the elimination of border restrictions on trade among members. Moreover, and partly because of the absence of common institutions, early RIAs excluded coordination of activities such as standards or government procurement which have continued to fragment markets. Thus, deep RIAs involving agruments on institutional deepening or on widening their scope can reduce market frogmentation.

## 'Deep' integration: what makes it effective against 'policy backsliding'?

As we have noted above, the effectiveness of RIAs is undoubtedly related to the policy objectives these arrangements are expected to target. In choosing a particular model of integration, therefore, one has to be clear about the objective of the RIA and what instrument will have be applied to achieve that end. As we have also noted, the effectiveness has been traditionally assessed in terms of welfare implications (gains or losses). In this respect, prevention of policy backsliding is not a usual element in the objective function of economists. However, if we were to judge the success of international agreements by the effectiveness with which they can prevent 'policy backsliding', a number of conditions would have to be satisfied. These conditions are both economic and political. This rest of this chapter will deal with the former.[37]

When the objective of policy makers is a stable trade policy, the choice of policy instruments is not immediately straightforward. On the surface, it may seem useful to distinguish between two sources of unstable trade policy – trade imbalances and special commercial interests. It is for these reasons that economists often make a distinction about policies that need to be taken to address each of these two issues. When pressure for changes in trade policy comes from special interests, so goes the argument, these pressures must be contained by the existing legal or political mechanisms.[38] When the pressure is due to trade imbalances, the solution is macroeconomic adjustment, not changes in trade policy.[39]

A deeper analysis shows that even the pressure of lobbies and other special interest groups often originates in economic inefficiencies and market distortions that can be corrected by policy changes. For example, as Feldstein (2000) argues, a single monetary rule for a monetary union such as the EU may not be suitable because cyclical and inflation conditions may vary substantially among members and because labour market conditions may also vary. These tensions may lead to protectionism. Similarly, an across-the-board and persistent loss of market shares of industry may be due to problems in the financial sector, such as poor design and implementation of prudential requirements or undercapitalization of banks. The resolution can only be made through appropriate *structural* policies that target the financial sector. Now, would these issues be better addressed with the help of a 'deep' RIA? Quite possibly – if the 'deep' RIA in question facilitates the adoption of better prudential rules, crossborder mergers and acquisitions, and involves financial and technical assistance in the example above.

Thus, to maintain stable trade policy, at least four conditions are necessary, none of which is a feature of 'shallow' RIAs. These conditions are (1) coordination of macroeconomic policy, including the exchange rate policy; (2) harmonization of product standards; (3) coordination of competition policies; and (4) a mechanism for the enforcement of trade disciplines.[40]

### Critical role of macroeconomic policy

Arguably, one of the most underrated policy areas that can make a difference to the success of a deep RIA is macroeconomic policy. A good macroeconomic policy is critical for trade as much as trade is typically very important for macroeconomic performance. This is a matter of simple economics. As shown in Drabek (2004), stability of trade policy is closely linked to the evolution of trade deficits and surpluses, since the latter are closely related to the evolution of the difference between national savings and investment. The link makes it particularly important for the management of balance of payments, for changes in trade balance are a determinant of the overall balance of payments. The dependence of balance of payments on the evolution of trade balance is particularly strong in countries with foreign currency restrictions on the capital account. In such cases, the evolution of a trade balance is critical. But even with free capital movements, the conduct of macroeconomic policy will be important. It will affect not only trade flows, but also capital movements in and out of the country and thus the net changes in the stock of national savings and investment. In sum, measures of macroeconomic policy will be always required to restore external balance.[41] Trade policy should only be used in extreme situations of balance of payments disequilibria.[42]

### *Coordination of macroeconomic policies*

What kind of macroeconomic policy should be pursued by countries if they form a regional grouping? What macroeconomic policies are likely to make a RIA effective, stable and, ultimately, more successful? The theory is clear about the choices that can be made by governments.[43] Macroeconomic policy based on discretion may give regional partners more flexibility but it is also likely to introduce less dynamic consistency in policy making among the regional partners. This, in turn, is likely to transmit imbalances from one country to another within a given RIA. (Fisher 1990). By implication, rules involving a coordination of macroeconomic policy are valuable to countries if they want to benefit from greater stability of trade policy. Coordination will be required in order to avoid unsustainable external imbalances. Moreover, the need for coordination of macroeconomic policies is likely to increase as countries become more interdependent – a typical situation of successful RIA (Genberg and de Simone 1993).

The theoretical conclusions are also supported by empirical evidence. A number of reviews of various regional trading arrangements from across a broad range of countries around the world and different types of RIA show that poor conduct of macroeconomic policies has typically been one of the critical reasons for failures of many of these RIAs. Misalignment of exchange rates, which led to significant undervaluation or overvaluation of currencies, jeopardized further trade liberalization or led to increased protection. Exchange rate misalignments were also hampering a smooth functioning of the clearing and payments systems. High and variable rates of inflation increased business risk and thus impeded trade. Furthermore, inflation is highly detrimental to the development and integration of financial markets.[44]

This coordination is more likely to come from regional as opposed to multilateral/global agreements. Examples of regional arrangements with extensive coordination of macroeconomic policies are currency unions and monetary unions, as noted above. The benefits from these unions are not only in terms of real economic integration and trade expansion, as we have also already noted. The benefits also come from other properties of currency unions such as those that originate in market flexibility. For example, the likelihood of effective pass-through of changes in external prices to the relative prices of tradeables – a condition for the formation of optimum currency areas – has been found to be quite high in currency unions (Corsetti and Pesenti 2002).

In more general terms, certain conditions will have to be satisfied in order to make currency unions effective. Frankel and Rose (2000) argue that the critical conditions for a successful currency union are trade intensity between the regional partners and the extent to which business cycles are synchronized. But this is where the consensus among economists ends. While some economists believe that business cycles must be symmetrical if the currency union is to survive, others, such as Carre, Levasseur and Portier (1996) and Imbs (1998), believe that symmetry is not critical. Ritschl and Wolf (2003) go even further when they test the importance of the initial level of trade on the formation of currency areas. While recognizing that currency union formation is beneficial for trade creation, they argue that currency unions tend to be established by countries whose initial trade was high. In other words, currency unions are endogenously determined, and trade creation follows only if countries operate in the optimum currency area.

### Harmonization of standards

As we have argued above, governments have increasingly turned to non-tariff instruments as vehicles of 'new' protection following the dramatic reduction of tariffs and other border restrictions on trade in most countries. Primarily, these instruments include various technical, sanitary and phytosanitary regulations, anti-dumping duties and other sectoral safeguards and preferential rules of origin. In addition, governments have also been known to restrict trade through regulations affecting public procurement contracts. In their review of partial equilibrium studies looking at the effects of food safety standards on trade flows, Maskus and Wilson (2001) conclude that products standards are equivalent to a level of protection greater than the legislated tariff rates. Messerlin (2001) reaches a similar conclusion with respect to the effects of SPS regulations and technical standards in manufactures on the EU trade flows. In general, the problem is that compliance with standards increases costs of production and a lack of harmonization of standards increases the costs even further with adverse implications for trade.

Thus, there is a perception based on economic theory that harmonization of technical standards and sanitary and phytosanitary standards will reduce trade barriers. If standards requirements also exist within an RIA, their harmonization should increase *intra-trade* within the RIA in question. Empirical evidence is, however, is only beginning to emerge. Fairly simple and preliminary econometric experiments were carried out for the European Union in the WTO and they are consistent with the above conclusion (Piermartini, 2004).[45] A strong theoretical case can also be made that harmonization of product standards will improve market access of third countries into an RIA, even though the outcome may still depend on the specific circumstances of a given RIA. Unfortunately, there is even less empirical evidence that would provide rigorous tests of these theoretical arguments.

### Coordination of competition policies

If 'new' protection justified on the grounds of different product standards is to be minimized, or perhaps even avoided, through the harmonization of standards, protectionism originating in public procurement, anti-dumping and rules of origin can be mitigated through cooperation in the area of competition policies. Public procurement contracts are an obvious example of how government conditions can affect the competition for public contracts. It should be the task of competition authorities to deal with any violations of competition rules in this respect. Predatory pricing, price fixing or any other forms of non-competitive practice, which are often behind the justifications for anti-dumping remedies, can be addressed through competition rules and overseen by competition authorities. However, a coordination of competition policies will at best resolve the problems of crossborder intra-trade and will not deal with similar problems of the members' external trade.

In reality, a coordination of competition policies in RIAs is an exception rather than a rule. As Hoekman (1998) shows in his review of several RIAs, common anti-trust disciplines are not necessarily created by members, even to eliminate the existence of anti-dumping. Members may have to set these disciplines for other, strategic reasons. However, he concludes that 'common disciplines affecting competition between firms in each jurisdiction may be needed to completely remove the reach of contingent protection'.[46]

As we shall see in more detail in the final section of this chapter, the best example of coordinated competition policies in a regional bloc setting is the European Union. The EU closely coordinates policies that directly affect competition among member states. Disputes that may arise in the area of competition are resolved through the activities of the EU competition authority. Moreover, conflicts, such as those arising from mergers and acquisitions, are less likely to occur in deep integration arrangements than in shallow ones (Neven and Röller 2000). Moreover, since the EU is also a political union, effective penalties for non-compliance may include different forms of political and peer pressures.

## Dispute settlement mechanism

Finally, what has made the WTO an extremely powerful institution is arguably the presence of an effective dispute settlement mechanism. If members of an RIA want to emulate the success of the WTO they will, therefore, also need to emulate the effectiveness of the WTO's dispute settlement mechanism. Once again, there are not many regional arrangements that have gone further than the WTO in resolving disputes. All RIA dispute settlement systems suffer from the handicap that they are only applicable to member countries, and therefore exclude disputes between members and non-members.

An effective dispute settlement mechanism is not *a priori* the optimal instrument of trade policy. As Bagwell and Staiger (1999) show in their model of regional and multilateral negotiations, the threat of future retaliations may not be sufficient to deliver the most efficient multilateral agreement. However, as they point out, an enforcement mechanism of a 'deep' RIA can play a positive role in multilateral negotiations. A properly structured ('non-shallow') RIA will lead to an improvement in multilateral tariff cooperation.[47]

An effective dispute settlement starts with a law establishing disciplines to prevent 'trade policy backsliding' and with an effective law enforcement mechanism. Both can be enhanced by regional cooperation. The adoption of legal disciplines in a trade area common to all members, as is the case with members of the EU, is a gigantic step forward. Effective adoption of common legal rules is facilitated in the EU by the fact that this aspect of legal harmonization is set as a pre-condition of membership. Moreover, effective implementation of legal rules is boosted by legal aid, which is seen as key to optimal law enforcement (Garoupa and Stephen 2003).[48]

Furthermore, 'deep' integration is likely to be more effective in *enforcing common external policy* of its members. As already noted above, 'deep'

RIAs are likely to dilute the political powers of individual members, disrupt the formation of rent-seeking interest groups and stimulate more efficient outcomes by seeking compromises among members with a different set of priorities.[49] Moreover, reviews of dispute settlement mechanisms in different RIAs suggest that these bodies can achieve consensus more easily than in a multilateral setting, that it is possible to design a speedier procedure in an RIA and other procedures that are amenable to amicable resolution of disputes.[50]

## Policy backsliding: a framework for assessing the impact of regional models

In the previous sections we have addressed the question as to whether regional integration arrangements can be conducive to the stable conduct of trade policy in general terms. In discussing these issues, we have primarily relied on the existing literature. In this section we shall take a more detailed look at three distinctly different models – 'shallow' RIAs, the WTO system and 'deep' RIAs, one which we call, for the sake of brevity, the 'WTO-Plus' model. The simplest model – the 'shallow' RIA – refers to free trade arrangements in which parties agree to coordinate the border measures of their trade policies. The second model is the multilateral trade arrangements under the WTO umbrella. Finally, the third model includes commitments that go beyond the WTO commitments. The models are summarized in Table 1.1 . To simplify our task, NAFTA will be used as a point of reference for the 'shallow' model.[51]

We shall be asking the following questions:

- Do regional trade agreements restrain governments from taking *additional* trade restrictive measures and thus promote stable trade policies?
- Are RIAs more or less conducive to the stability of trade policy than the multilateral agreements under the WTO?
- What are the reasons that some RIAs seem to be more effective in this respect than others?

We shall hypothesize that the answer to these questions will depend on the type of RIA. In order to make an assessment, we shall apply the following methodology. We shall assume that the effectiveness of each of these three models depends on (1) the coverage of the agreement; (2) on the way the agreement is structured and on the types of instruments that are permitted and/or stipulated in the agreement; (3) on the way the agreement is implemented; and (4) on the way the agreement provisions are enforced.[52] The term 'coverage' of agreement refers to the coverage of sectors and economic activities. Measures permitted under the agreement refer to those measures

*Table 1.1* Provisions in trade agreements: WTO, NAFTA and the European Union

| Issue | WTO | NAFTA | European Union |
|---|---|---|---|
| **Structure** | WTO administers several agreements | System comprises bilateral agreements among US, Canada and Mexico. Some side agreements and understanding | Wide-ranging set of laws under *Acquis Communautaire*. The provisions leading to significant deepening of the European Community were incorporated in *Single Market Act* with further measures adopted in the Maastricht (1992) and Amsterdam (1999) Treaties. Establishment of a broad range of institutions and of the European Monetary Union (1.1.1999) and common currency |
| **General principles** | Non-discrimination National Treatment | Non-discrimination National Treatment | Non-discrimination National Treatment |
| **Exceptions** | General exceptions for security, health and safety, quarantine and moral issues | Follows WTO | Follows WTO |
| **Rules of origin** | Loose guidelines. Work program to harmonize non-preferential ROO | Stringent rules, particularly for clothing, textiles and automobiles. High value added required (about 50–60 per cent) | Complex rules (varying values of value added required), high administrative costs due to extensive use of *ad valorem* percentage criterion, and stringent rules particularly for clothing and textiles. Applies both preferential and non-preferential ROO |
| **Tariff reductions** | As determined in negotiating rounds | General goal of phase out = immediate, or in 5; 10; or 15 year programms. Mostly 10 years. Some exceptions | No border restrictions for members. As determined in negotiations for non-members |

*Continued overleaf*

Table 1.1 Continued

| Issue | WTO | NAFTA | European Union |
|---|---|---|---|
| **Non-tariff measures** | Quotas prohibited except in some specified circumstances | Quotas prohibited except in some specified circumstances | Quotas prohibited except in some specified circumstances (textiles and clothing to be eliminated 1.1.2005, and certain products from China. Also special regimes for imports from several transition. Quotas on imports from Belarus, Uzbekistan and Viet Nam, and quotas on steel from the Kazakhstan, Russia and the Ukraine. QR on imports from Outward processing Traffic except for approved situations. Measures to protect the consumer, environment and animals, to enhance domestic and foreign security |
| **Duty drawbacks, remissions on duty** | Not illegal under GATT/WTO | Gradual elimination of duty drawback | Not applicable in trade within the EU due to absence of duties. The Commission retains the competence to grant duty drawbacks in trade with non-members. Rebates of value added on exports from the EU exist |
| **Agriculture** | Export subsidies permitted within limits, high domestic subsidies permitted, quotas apply for limited access in some products. | New reductions to be negotiated. Secured free access to the US market, some products (corn, sugar, orange juice, vegetables) have longer transition periods (of up to 15 years). Canada will apply WTO rules in trade with Mexico and US | Agriculture is subject to complex set of rules under *Common Agricultural Policy*. The policy covers production and trade of most agricultural products and applies instruments of domestic production support, export subsidies and import restrictions. The 2003 CAP Reform removed the link between subsidies and production ('decoupling') even though individual Member States may choose to provide some of the original support (partial 'decoupling') |

*Table 1.1* Continued

| | | | |
|---|---|---|---|
| **Automotive** | No special rules | Progressive removal of restrictions on trade and investment controls between Canada and US with Mexico for periods up to 25 years | No special rules. Measures to increase competition in the distribution network (see further below) |
| **Clothing and textiles** | ATC terminated at end of December 2004 All textile and clothing products integrated into the WTO rules thereafter. Special safeguards rules apply | Commitment to remove barriers over a 10 year period. Exceptions permitted. Special safeguards rules apply | Commitment to remove quotas by 1.1. 2005 |
| **Energy and petrochemicals** | No specific provisions | Specific rules to secure trade rights and to govern actions of regulatory bodies | Special provisions public supplies, works, services and remedies and governments procurement. (concerns also water, transport and telecommunications – utilities directive) |
| **Subsidies** | On industrial products, export subsidies are banned, certain domestic subsidies are actionable, all subsidies are subject to countervailing action. Looser rules apply to agriculture subsidies. GATS – national treatment applies. Specific rules under negotiations | Export subsidies are banned in agricultural trade between Canada and the USA and circumscribed in agricultural trade with Mexico | Export subsidies are banned within the EU. Any new production subsidies are subject to approval of the Commission. Pledge to shift subsidies from supporting individual companies or sectors towards tackling horizontal objectives (e.g. employment, environment, regional development, R&D) |

*Continued overleaf*

Table 1.1 Continued

| Issue | WTO | NAFTA | European Union |
|---|---|---|---|
| **Anti-dumping and counter-vailing** | Duties can be imposed if damage by subsidized or dumped imports can be demonstrated. No specific rules in GATS | Rights to challenge and review countervailing and anti-dumping actions are provided | No anti-dumping and countervailing provisions against imports from Members. AD and CVD against other countries permitted and governed by WTO rules and in the competency of the Commission |
| **Safeguards** | Rules stipulate when members may restrict imports causing damage. In GATS rules under negotiations | Rights are provided to take safeguard action in trade within NAFTA. NAFTA members have rights to be exempted from safeguard actions applied against countries outside NAFTA | No safeguards in trade within the EU. Trade with non-members governed by WTO rules such as Art. XIX of GATT and the relevant WTO Agreements and the special mechanism of the WTO Agreement on Agriculture. The EU also applies a special safeguard regime in trade with China and another with third countries, not WTO Members. Competence in the Commission |
| **Government procurement** | Government procurement excluded from GATT rules, but a plurilateral agreement limits this exclusion | Rules restrict favouritism for national suppliers and require national treatment and transparency | Rules restrict favoritism for national suppliers and require national treatment and transparency. The coverage is somewhat bigger and description procedural rules more detailed than in NAFTA. Harmonization of directives for members states legislation, regulation to mandate the use of 'Common Procurement Vocabulary', standard forms contract notices. Special provisions in utilities |
| **Sanitary and phytosanitary measures** | The right to restrict imports on quarantine and health grounds is restricted to justifications based on sound science and risk assessments | Provision reflect WTO rules and scientific basis but more focus on rights of governments to protect, less on obligation to justify. | Also espouses right to restrict trade where scientific justification exists but greater recognition of precautionary principle |

*Table 1.1* Continued

| | | | |
|---|---|---|---|
| **Standard and technical barriers** | Standards and mandatory technical regulations must be nondiscriminatory and provide national treatment and be based on sound science | Rules are similar to WTO, but more permissive in some respects when consumer and environment issues are in play | Rules are similar to WTO. Moreover, considerable efforts to harmonize standards, technical regulations and conformity assessment procedures as well as for the acceptance of conformity assessment results. Products placed into the EU market must comply with relevant regulations, where they exist to meet health, safety and environmental objectives, established by means of conformity assessment procedures. In all other cases, general product safety requirement applies. Relevant regulations also exist regarding advertising, contractual arrangements, defective products |
| **Customs administration** | Rules govern customs valuation. Agreements on import licensing and on ROO also apply | Rules govern administration of rules of origin | Rules govern customs valuation, import licensing, ROO, The administration also covers other areas such as the Community Customs Code, movement of goods into and from the Community, admission of goods into the EU market, customs clearance, harmonization of VAT requirements for invoicing, cooperation to combat fraud |
| **Competition policy** | No specific agreement. *Ad hoc* provisions on competition-related issues contained in the GATS, TRIPS and other agreements. A set of legally binding principles for telecommunications have been annexed to | General commitments to maintain provisions dealing with anti-competitive practices and to implement related cooperation agreements (not subject to dispute settlement). Also enforceable provisions | |

*Continued overleaf*

Table 1.1 Continued

| Issue | WTO | NAFTA | European Union |
|---|---|---|---|
| | individual GATS Schedules by many Members | targetting anti-competitive behavior of state enterprises and monopolies. Anti-dumping remains applicable to internal trade flows (!) | Treaty provisions cover all forms of anti-competitive behaviour and abuses of dominant position in addition to anti-competitive subsidies by Member States and state monopolies of a commercial character. Community regulation dealing with anti-competitive mergers. Block exemption regulation targeted for motor-vehicle distribution and servicing agreements |
| **Temporary entry** | No specific provisions. Scope for temporary entry of natural persons in the GATS | Provisions lay down rules for temporary entry for business purposes | Provisions lay down rules for temporary entry based on ATA Convention and the Istanbul Convention. The ATA carnets are issued under the responsibility of the national guaranteeing associations |
| **Services** | GATS employs negotiations to Additional provisions in progressively liberalize market access and national treatment, only applicable to sectors specifically identified by Members in their Schedules ('positive' list approach). MFN required as a general obligation. Annexes apply to financial services, | Aim is to liberalize cross-border trade in non-financial services. Core liberalization principles are MFN and national treatment (obligation to provide the best of the two). The Chapter abolishes local presence requirements. Non-conforming mesures must be listed in specific annexes. ('negative' list approach) | Free movement of services is one of the Four Fundamental Freedoms of the single market. The goal was to achieve fully integrated securities market by 2003 and financial markets by 2005 and to foster 'the information society' through enhanced competition for telecommunication services and a new framework for e-commerce |

*Table 1.1* Continued

|  |  |  |  |
|---|---|---|---|
|  | telecommunications, labour and air and maritime transport | Canada excluded cultural services, the US excluded maritime transport and government services, including health and social services. No residency restrictions are permitted on licensed services providers | |
| **Tele-communications** | An Annex to GATS sets out rights of access to networks made available to the public. The Information Technology Agreement requires elimination of tariffs on most IT products. Safeguards against anti-competitive practices provided for in the Reference Paper. | Tariffs on Telecoms equipment removed over 10 years. Access must be provided to public networks, with exemptions on public interest grounds licensing, universal service obligations, privacy, etc.) | Since 1.1.1998 all telecom infrastructure and service segments have been open to competition. The framework has been extended to other foreign operators under the Fourth Protocol to the GATS with minor limitations on market access and NT. Harmonization of the framework for national legislation on licensing and access to the network, on local loop unbundling, on electronic communication (interconnection, |
| **Financial services** | GATS principles apply in general to financial services. An Annex to the GATS details provision for financial services, including the possibility of recognizing other Members' | A separate chapter applies to cross-border trade in financial services and the establishment of financial institutions. Provides for liberalization of cross-border trade; right | Liberalization is based on three principles: (1) minimum harmonization of standards at the EU level, (2) mutual recognition of regulations among EU states; and (3) home country control (i.e. supervision by the EU country of registration) EU institutions and foreign subsidiaries have the right to operate in all countries when they are registered in just one |

*Continued overleaf*

50

Table 1.1 Continued

| Issue | WTO | NAFTA | European Union |
|---|---|---|---|
| | prudential regulations, and an exception to safeguard the adoption of prudential rules | of establishment; national treatment and MFN for foreign institutions and investors. Exceptions to these principles can be scheduled in specific annexes at the time of entry into force. National treatment is qualified by a right for reciprocal treatment | country ('single passport' rule) |
| **Investment** | Covered in WTO only by GATS, where progressive liberalization of commercial presence is negotiated, and in limited respects by the TRIMS TRIPS and Public Procurement | Foreign investors get national treatment and a right of nondiscrimination in relation to establishment, conduct, acquisition, expansion and management of investments. Conditions on investment are generally not permitted. Exemptions apply. Investors have a right to establish. A tribunal can settle disputes between investors and Governments | Free movement of capital is one of the Four Fundamental Freedoms of the Single European Act within the Community. Since 1996 investment firms have benefited from the single passport under the 1993 Investment Services Directive, with further policy modifications under way – to target insider trading and market manipulation, provision for single prospectus, extension to mutual funds, Community-wide framework for collateral, revision of disclosure requirements |

*Table 1.1* Continued

| | | |
|---|---|---|
| **Intellectual property** | TRIPs Agreements sets out minimum standards for copyright, industrial property, trademarks and integrated circuits and rules on geographical indicators, lays down requirements for their enforcements and makes these obligations subject to the WTO disputes settlement procedures | NAFTA rules are similar to WTO rules | A greater level of harmonization has been attained in a number of areas through various legal instruments e.g. resale author rights of original work of art, copyrights and related rights for the digital environment. Attempt to strengthen the enforcement of IPR by a harmonizing directive to go beyond the 'minimalist approach' of the TRIPS Agreement |
| **Sub-national government** | WTO obligations fall on national governments which are responsible for compliance by subnational governments | Central governments are obliged to ensure sub-national entities:<br>• apply national treatment on services and investment issues;<br>• apply NAFTA rules on regulation of financial services;<br>• treat foreign investors without discrimination | |
| **Dispute settlement** | WTO disputes procedures amount to compulsory arbitration | Legally binding dispute mechanism established. However, importers may choose to petition to | Legally binding dispute settlement established as well as the European Court of Justice. The latter covers not only 'technical' disputes but also broader issues |

*Continued overleaf*

Table 1.1 Continued

| Issue | WTO | NAFTA | European Union |
|---|---|---|---|
| **Labour and environment** | The exemptions provisions of GATT permit most environmental restrictions | different jurisdiction – domestic, NAFTA and the WTO | Free movement of labour is one of 'Four Fundamental Freedoms' of the European Single Market. Environmental policy incorporated into the EU through institutional mechanism of the Community, and was only later incorporated in an intergovernmental treaty (Single European ACT of 1986). Like NAFTA, the trade-environment rules are more developed than in the WTO |
| | | No explicit environment or labour provisions. NAFTA obliges members to give priority to environment agreements where provisions clash with other agreements. Environmental agreement was negotiated alongside the trade agreement | |
| | GATS Mode 4 provides for progressive liberalization of natural persons who are supplying services and those who are employed by service providers. The GATS Annex on Movement of Natural Persons provides clarification of the scope of mode 4 | Separate side agreements on labor and environment deal with non-trade issues among NAFTA members | |

*Sources*: Eliassen and Borve Monsen (2001); Geradin and Kerf (2003); Hoekman (1998); Spragia (2001); Telo (2001b); 'The Australia – USA Free Trade Agreement: Issues and Implications', Australian APEC Study Centre (2001); WTO (2002b).

that allow the imposition of trade restrictions. The implementation process refers to the way in which these restrictive measures are implemented. Finally, the enforcement process refers to the way in which the trade restrictive measures are enforced.

In order to assess the capacity of international agreements to contain protectionist pressures, it is necessary to identify the origins of protectionist pressures. As explained above, these origins can be grouped under two headings: macroeconomic disequilibrium and sector-specific pressures (interest lobbies). Macroeconomic pressures for protection can, in turn, be contained either by a coordination of macroeconomic policies or, in the absence of coordination, by special safeguard provisions such as the balance of payments provisions of Article XVII of GATT. Sector specific pressures can be contained by domestic policies, such as competition policies or other government measures, to remove distortions in particular markets. Alternatively, agreements will normally contain safeguard provisions against subsidized exports, dumping and surges of specific imports.

### 'Shallow' RIAs

There has been a major proliferation of RIAs in recent years. As of June 2002, the GATT/WTO Members notified 254 RIAs. Only Hong Kong, China, Macau and Mongolia, among WTO Members, did not participate in RIAs. Of the 254 RIAs notified to GATT/WTO from 1947 to June 2002, 129 were notified between January 1995 and June 2002.[53] By far the greatest number of these RIAs were free trade agreements or customs union agreements, with bilateral free trade agreements clearly dominating. The main feature of these agreements is the reduction or complete elimination of border restrictions on trade among regional partners and, in the case of customs union, harmonization of trade policy vis-à-vis third countries.

The agreements tend to be limited in coverage. While they have been effective in reducing or eliminating tariffs on industrial goods, they have rarely targeted agriculture. Being primarily confined to border measures, these RIAs do not typically address trade distortions resulting from inside-the border-measures, such as differences in regulations, the presence of anti-competitive practices and so on. Although a free trade agreement in name, NAFTA goes further than a 'standard' free trade agreement (Table 1.1, column 3). It also provides for movement of capital (investment) and, to some extent, for movement of labour and protection of environment. Even though NAFTA is presented here as a 'shallow' RIA, these provisions make it somewhat wider than the WTO in these areas.[54]

Most importantly, no attempt is made to coordinate domestic macroeconomic policies. Moreover, unlike the WTO agreements, NAFTA and other free trade agreements typically provide disciplines for preferential rules of origins. This is a serious issue under free trade agreements since they can be extremely costly, especially for small developing countries that often conclude regional arrangements with the view of obtaining preferential market access.[55]

The second most notable characteristic of NAFTA is the provision allowing members to apply anti-dumping measures. This limits the effectiveness of NAFTA to contain pressures for new protectionism. Anti-dumping measures have been used in NAFTA both in agriculture and in manufacturing sectors. Indeed, anti-dumping has been the most frequently used safeguard instrument by member states, and it has been widely seen as 'trade-deflecting' and as an instrument of protection.[56] No common rules on competition policy are part of the agreement with the exception of anti-competitive practices of state enterprises, even though a proposal was made to discipline anti-dumping and to link anti-dumping with anti-trust (Hoekman 1998).

## WTO 'model'

The WTO 'model' and its scope for new protectionist measures is also summarized in column 2 of Table 1.1. In the WTO, the new trade restrictions of members can legally arise in several forms. As we have noted above, the best known instrument safeguarding against future tariff increases is *tariff binding*. The instrument of tariff binding refers to the level of tariff that the country imposing the tariff agrees not to exceed. The bound tariff must be contrasted with actual tariffs; that is, tariffs that are actually applied by countries and which can be legally increased up to the level of the binding. If tariffs are not bound in the WTO, the country concerned may increase tariffs despite the limitation of the binding rule. In brief, the concept of binding gives countries the opportunity to set their actual tariffs at a level below their bound commitments in the WTO. Moreover, developing countries have not even been requested to bind 100 per cent of their tariff lines, as is typically the case of advanced countries. In other words, the WTO agreements allow 'backsliding' of tariff policy without actually violating the WTO commitments.

The critical question is, therefore, whether countries have actually 'bound' their tariffs, and whether actual tariffs are identical or lower than the corresponding bound tariffs. This is an empirical question. Many developing countries, for example, apply tariffs that are far below the corresponding tariffs bound by these countries in the WTO agreements. Moreover, the level of binding tends to be lower than that of developed countries, as already noted. Most developed countries bind all their tariffs, which means that the possibility of increasing actual tariffs up to the level of the corresponding bound tariffs does not exist.

The WTO also allows for other trade restrictive measures. As with 'shallow' RIAs, these include anti-dumping responsibilities as remedies against anti-dumping, and a variety of safeguard measures as well as measures to protect domestic markets against import surges. Furthermore, the agreements also allow countries to impose trade restrictive measures at times of balance of payments difficulties. The enforcement of trade disciplines takes place in the WTO through the well-known dispute settlement mechanism.

The WTO Agreements outlined above have other shortcomings that are relevant in assessing the scope for trade restrictive measures. Briefly, (1) the coverage of sectors remains limited. This means that governments can pursue whatever policies they choose in those sectors in which they have made no commitments without violating any of their international obligations. (2) The Agreements contain many exceptions. In negotiating GATS, members agreed on the negotiating modality 'request-offer', which only targeted sectors that the members were willing to offer in their schedules. Thus, services were only liberalized in those service sectors that appeared on the list. Once again, the government policy in the rest of the service sectors is at its discretion. (3) Not all countries have been obliged to 'bind' all their commitments, thus giving rise to legal loopholes and to legal arguments justifying increased protection.[57] (4) The application of operational safeguards has often been criticized as an instrument of protection from foreign competition rather than a temporary relief to provide room for the adoption of restructuring measures or as a true safeguard against unfair competition. Moreover, safeguard measures have been extended in real situations well past the original deadlines. (5) Last, but not least, despite the expanded coverage of topics for negotiations, the negotiated agenda remains limited. For example, important economic activities such as foreign investment are not part of the Agreements, except in a number of well-defined areas – most notably GATS, TRIMS, and TRIPs. This, too, can adversely affect trade since a growing share of foreign direct investment is closely linked with trade due to their complementarities.[58]

### 'Deep' RIAs ('WTO-Plus')

We shall now briefly consider our third model of international cooperation – a model of 'deep' regional integration. By way of an example, we shall refer to the legal framework of the European Union. The EU model has all the basic features of the WTO and, in addition, it contains extra elements in excess of the WTO rules and disciplines. For this reason, we have called the model 'WTO-Plus'.

As a 'blueprint for the model', let us consider the Europe Agreements – the agreements of Central and East European countries with the European Union. The choice of Europe Agreements is, of course, not accidental. The agreements refer to countries whose trade policies have been subject to considerable scrutiny, since these are countries acceding to the European Union.[59] Moreover, the model provides a convenient example of fairly far-reaching 'deep' integration and, once again, the model is summarized in Table 1.1, column 4.

The main features of the Europe Agreements are: first, in contrast to WTO provisions, the agreements act as a forerunner to the ambitious objective of these countries to accede to the European Union and integrate economically as well as legally, politically, institutionally with the existing EU members. The Agreements provided for cooperation between the EU and the Associate Countries of Central and Eastern Europe in different fields include legal approximation, political dialogue, advice on democratization and political processes, environmental, transport and customs policies, institution building to support market-type economies.

Second, the Europe Agreements and the subsequent Accession Agreements contain a number of provisions that go well beyond the existing disciplines of the WTO. These extra provisions range from harmonization of standards on technical and food safety to procedures as well as standards in the area of customs administration. Poland, for example, had adopted 30 per cent of the total number of technical standards by June 2000. This number increased to 45 per cent by June 2001, and reached 80 per cent by the time of accession in 2004.[60] By adopting the EU standards through the Europe Agreements, the Central and East European countries clearly make deeper commitments than they have made in the WTO, TBT and SPS Agreements, as well as in the Agreement on Customs Administration.[61]

Third, the associate countries have agreed to adopt the EU standards and rules for public procurement. Fourth, the economic coverage of the Europe Agreements is far greater than that of the WTO agreements. In particular, the Europe Agreements include fairly detailed provisions for cooperation in the area of foreign investment which, it will be recalled, is not subject to a comprehensive investment agreement in the WTO. The coverage of service sectors is also much more extensive in the Europe Agreements than in the WTO. By way of examples, the Europe Agreements – in contrast to the WTO – make provisions for real estate related services, and much greater commitments in financial services, legal services or ownership, use, sale and rent of real estate.[62] Fifth, although modest in their scope, the Europe Agreements make provisions for movement of labour but, of course, with the provision for full labour mobility after the accession following transition periods.

Sixth, what makes the cooperative arrangements between the Central European countries and the European Union particularly interesting and different from the WTO Agreements is the clear attempt of the former to coordinate their macroeconomic policies with those of the EU. Although, our RIA WTO-Plus model makes no formal provision for such a coordination, all the acceding countries have been making considerable efforts to harmonize their monetary policies to those of the EU. In the early 1990s, when the Central European countries were signing the Europe Agreements, the EU was a single market without a monetary union but with close monetary cooperation. The harmonization of monetary policies was first reflected in the policies to stabilize the exchange mechanism by creating ('shadow') exchange rate pegs supported by exchange rate bands in all participating countries.

Monetary cooperation has been one important reason why protectionist measures have been, by and large, absent in the trade relations between these acceding countries and the EU. Monetary cooperation has reduced exchange rate volatility, which is detrimental to exports and domestic production. The empirical evidence from the literature from outside the region is quite clear, and it shows that a low or moderate inflation reduces macroeconomic distortions, such as overvalued exchange rates, which have been character- istic of fast-growing economies.[63] Thus, maintaining balance of payments equilibrium became a matter for EU member countries of, at first, adopt- ing monetary rules and, subsequently, of inflation targeting, establishing an exchange rate anchor, attempts to introduce fiscal discipline and to control the overall level of debt needed to maintain the stability of exchange rates of the members' currencies. In the extreme case of excessive BOP pressures, which would be reflected in the pressure on the country's exchange rate, the rate itself may have to be adjusted – but still without recourse to additional restrictions on imports.

These policies of close monetary coordination were also increasingly used by the acceding Central and East European countries. Initially, in the early 1990s, the countries' macroeconomic policies were unstable, resulting in bal- ance of payments pressures, and this led to repeated steps to curb imports to protect the balance of payments.[64] Indeed, several of these countries invoked exemptions under the WTO provisions and applied temporary import restric- tions for balance of payments reasons. The measures typically led to the imposition of uniform import surcharges.[65]

Over time, however, their policies have changed. The early attempts to stabilize exchange rates in these countries were directed towards tying their currencies more closely to the EU currencies, and eventually to the euro itself. Subsequently, the policies have changed towards the harmonization of the countries' exchange rate policies with the EU through the introduction of monetary rules and inflation targeting, and more prudent rules on external borrowing.[66] Although not formally anchored into the Europe Agreements, the acceding countries' commitments to maintaining macroeconomic stabil- ity and to the convergence to the EU currencies have greatly reduced the pro- tectionist pressures in their countries.[67] In addition, the proximity to the EU markets – a factor leading to lower transaction costs – high intensity of trade, fairly symmetric business cycles and emerging labour mobility provided ele- ments for the region to become an optimum currency area in everything but name, thus making the conduct of monetary policies more effective.

The last, but not the least, important factor of effectiveness of different RIAs is related to *the level of tariff protection* that governments accord to their industries. In general, the link between tariff level and trade policy back- sliding is not straightforward and probably varies from country to country, depending on a variety of conditions. Nevertheless, it is very likely that high tariffs tend to reduce the pressures for *additional* protection and *vice versa*.

Moreover, the lower the level of tariffs and the higher the level of bindings, the smaller the role of border measures to restrain imports and the greater the importance of inside-the-border instruments. It follows, therefore, if inside-the-border measures (regulations) are fully harmonized with trade partners, the main instruments of 'protection' become macroeconomic policy, competition policy, and policy on state aid. This is essentially the current situation within the European Union, and was increasingly the pattern in the three countries studied in this project, the results of which studies are reported in the empirical chapters of this book.

As a result of the attempts to coordinate monetary policies and the increasing and widespread adoption of the EU technical, health, sanitary and other standards, the scope for the application of traditional instruments of trade policy – border measures – has been greatly reduced. Similarly, the recourse to anti-dumping measures has been eliminated in the EU and the restrictions can only be applied against non-members. *Pari passu*, anti-dumping responsibilities were only permitted to be applied by EU Members against the acceding countries until the accession.[68] In those cases, the WTO rules applied. In the meantime, the competition authorities in the acceding countries were targeting strategic objectives rather than trade competition whenever considering mergers and acquisitions.[69]

Similarly, countervailing measures and measures to protect members against import surges can only be applied against non-members and, for the duration of the Europe Agreement, the measures could also be applied against the acceding countries. Moreover, additional leeway was given to the acceding countries under the so-called Restructuring Clause (Art. 28).[70] Once again, in such cases the WTO rules applied.

Finally, the objectives of the Europe Agreements and the supporting policies were so powerful that they acted as extremely effective tools to enforce the acceding countries' trade and investment commitments. In contrast to most RIAs, the Europe Agreements and the subsequent accession negotiations have brought domestic reforms directly into the text of the Agreements.[71] As a result, these legal commitments have become incentives to avoid sanctions or other punishments. The latter operate partly through 'peer pressure' and partly through opportunity costs of non-compliance. The latter could be quite serious, as non-compliance could affect the access of acceding countries to 'structural funds', to funds for technical assistance or provoke other retaliations. The retaliations could be particularly costly due to the dependence of these countries on the EU market.

## Conclusion

Minimizing the risks, renewed protectionism requires a number of conditions that have to be satisfied. As we have argued in this chapter, the conditions include a closer macroeconomic cooperation, harmonization of standards, coordination of competition policies and an effective enforcement mechanism. None of these conditions are currently satisfied in the current system of WTO disciplines or in most regional agreements. 'Shallow' regional arrangements are poor instruments *sui generis* in this respect. Unfortunately, not even the WTO rules are optimal to prevent 'new' protection and 'backsliding' in trade policy. There still are also several legal instruments in the WTO agreements allowing members to use protective measures in specific circumstances. These measures are not the first-best instruments.

We have argued in this chapter that 'deep' regional arrangements can be very effective in containing a great deal of pressures for 'new' protection. An agreement on any of the conditions mentioned above is more likely to be reached among a small number of countries than among 147 or so members of the WTO. Moreover, an RIA is likely to put together countries that have an *a priori* better chance of signing an agreement once they decide to integrate more deeply with their partners. The effectiveness of the WTO system would also be greatly enhanced if appropriate changes in the coverage of the agreements and the actual disciplines were introduced.

'Deep' integration, if properly designed, can also be an instrument of net trade creation, as the available evidence tends to suggest. 'Deep' integration should lead to improved governance as members adopt common and more transparent rules on corporate behaviour. In addition, a 'deep' RIA that is effective against creeping 'new' protection among the trading partners will lead to a considerable reduction policy uncertainty. This, in turn, will result in lower risk premia among investors – an issue that should be of particular interest to developing countries, especially those that have a limited access both to global markets and to foreign investment.

'Deep' integration will not, however, address the remaining problems of regionalism, which can still be serious. In particular, these arrangements continue to rely on detailed and complicated rules of origin. These rules are not only discriminatory but also highly detrimental to trade incentives. Moreover, any given 'deep' RIA does not solve another fundamental problem of Rules of origin – the multiplicity of complex ROO and the costs of their implementation. The latter can only be resolved through a multilateral initiative such as in the WTO system.

## Notes

1 Some of the features listed below are not necessarily contingent on *free* trade *per se* but on trade in general. Nevertheless, by removing *all* restrictions on trade, the scope for benefits is arguably maximised.
2 For more details, see Crawford and Laird (1999).
3 The important role of geography as a determinant of regional trading arrangements has been emphasized by Scollay and Gilbert (2001). See also Greenaway and Milner (2002). However, as pointed out by Bhagwati and Panagariya, the concept of 'natural' trading partners is vague and reflects different ideas to different people. See Bhagwati and Panagariya (1999), pp. 56–64.
4 Hoekman and Konan (1999), p. 1.

5    See further discussion below.

6    In fending off critics of regionalism, Lawrence (1996) emphasizes the point, correctly in my view, that the proper comparisons are not between a preferential arrangement and a completely free trade but between second-best situations. The latter must typically include under present conditions both RIAs and multilateral liberalizations in the WTO.

7    An alternative approach was taken by Baldwin and Venables (1995), who make a distinction between three different effects of RIAs – *allocation effects* which come from trade diversion or trade creation, scale economies and imperfect competition; *accumulation effects* which come from increased returns to different kinds of capital, and *location effects* which come from geographical dispersion of firms' activities and their effects on income (in)equality between regions.

8    Deep integration can be conducive to enhancing economic and political stability even in countries which are inherently or deeply unstable. The recent role of the European Union in the Balkans is a case in point. The EU insistence and pursuit of its proposal for 'Stability Pact' which has led to a normalization of trade relations with a market liberalization among the countries of the former Yugoslavia and their relations with the European Union, has been no minor achievement. It is inconceivable that the same effect would have been achieved through any multilateral initiative or other bilateral or regional arrangements.

9    See Telo (2001a), p. 90.

10   The argument is somewhat more complicated, as shown by Bhagwati and Panagariya (1999). The welfare losses can come about not only from a diversion of imports from a more efficient supplier to a less efficient supplier, but also from a deterioration of terms of trade against the high tariff country as a result of formation of a RIA. In other words, countries may experience welfare loss even without a trade diversion. See Bhagwati and Panagariya (1999), pp. 38ff.

11   See further discussion below.

12   The conditions are discussed at length in Bhagwati and Panagariya (1999).

13   The Kemp-Wan proposition on welfare improving RIAs through endogenous tariff policy is rejected by Jagdish Bhagwati. He argues that even if the RIA in question is welfare-improving *in the short run*, the incentive structure within the union will prevent further liberalization. See Bhagwati and Panagariya (1999). This is clearly another empirical problem. Recent evidence from the negotiations of the US–SACU trade agreements would go towards supporting Bhagwati's proposition rather than those of 'deep integrationists'. See Whalley and Leith (2003) But the jury is clearly out on this issue.

14   The debate which was started by a seminal article by Krugman (1991) focused on the role of regional arrangements in the process of global, multilateral trade liberalization. But as Winters points out, the criticism of the optimal tariff literature stems primarily from the view that tariffs and other protectionist instruments are not the outcome of finely-tuned debates about optimality of protection but, rather, the result of political pressures and negotiations. See Winters (1999), p. 18.

15   For specific model with these characteristics, see Winters (1999), pp. 11–19.

16   The first systematic treatment of issues of fiscal federalism are attributed to Oates (1999).

17   See, for example, Perotti and von Thadden (2003).

18   See, for example, the WTO (1995) for a review of the debate.

19   The literature on the effects of the recent accessions is large with fairly uniform conclusions. See, for example, Neven and Wyplosz (1996). Further evidence is provided in the case studies of the Czech Republic, Hungary and Poland in this volume.

20 The econometric work of Rose has been sometimes criticized on a variety of grounds. The most serious criticism has been addressed by Peter Kenen (2002). Using an alternative definition of currency union and Rose methodology based on gravity models, Kenen re-estimates Rose's equations and finds that Rose's results survive. For more discussion and evidence, see also Drabek (2004).

21 As is well known, the theory of optimal currency areas stipulates the importance of four conditions: mobility of labour and capital as well as free movement of goods and services, synchronization of economic cycles, symmetry of external shocks and a fiscal coordination. These conditions, in turn, presuppose strong trade linkages that may result either from special regional arrangements or simply be given by the 'geography'.

22 However, see also further discussion below. A recent study by Sapsford *et al.* (2002) somewhat contradicts these conclusions.

23 For more details, see Bhagwati and Panagariya (1999)

24 However, see the argument of Panagariya and Dutta Gupta (2001) noted in the text above, which puts a rather different light on Krueger's forceful point.

25 The empirical literature of specific RIAs is extensive. The reader may consult, for example, Andersen and Blackhurst (1993), Bhagwati, Krishna and Panagariya (1999) or WTO (1995).

26 See, for example, de Melo *et al.* (1993).

27 See Bhagwati (1999) for more details.

28 Notwithstanding these criticisms, even the critics of regional initiatives such Cooper, Bhagwati, Winters and others recognize the successes of 'deep' integration arrangements, especially those of the European Union. See Eliassen and Monsen (2001), pp. 122ff.

29 For more details, see Crawford and Laird (2000), pp. 8ff.

30 The controversy was stirred with a paper by Summers (1999) and subsequently followed by Krugman (1999) which was hotly contested by Bhagwati and Panagariya (1999). On the policy level, the idea of 'natural' trading partners goes back much further. It is perhaps no exaggeration to say that each movement towards the establishment of a regional bloc has always raised this issue.

31 The example is not hypothetical. The choice between both blocs – the 'West' and the 'East' – was precisely what most transition countries faced after the collapse of central planning. See, for example, Sorsa (1994).

32 This goes a long way in the direction of Bhagwati and Panagariya's argument, which shows that a high *initial* level of trade is not a precondition for countries to be natural partners. The experience of transition countries provides powerful empirical evidence to their argument.

33 Similar results were obtained in a formal model by Venables (2003) for reasons discussed above in the text. He suggests, therefore, that RIAs of North–South type are more likely to succeed than those of the South–South variety.

34 For more details, see also Baldwin and Venables (1995).

35 The experience of transition countries with regional agreements in COMECON is a particular case in point. These countries tried to deepen their integration for almost fifty years but were unable to do so beyond rudimentary bilateral agreements on trade. Neither the economics nor politics of COMECON allowed it.

36 This has been one of the main points made by Winters (1993) about the EU enlargement. He argued that the EU enlargement – its timing and conditions – have been managed in such a way as to reduce the adjustment pressures on the incumbents.

37 This may be a great simplification of real constraints. The power of lobbies or legal provisions or simply the way societies are organized may certainly make a difference to the conduct of trade policy. Political scientists have, of course, their own explanations. The *neo-functualists* believe that the Maastricht Treaty of 1992 was the critical impulse strengthening supranational institutions and the responsibilities of the Union. See, for example, Bussiere and Mulder (1999) and Eliassen and Monsen (2001), p. 122.

38 I am assuming, of course, that the pressure comes from *domestic* lobbies. When the pressure comes from foreign lobbies, the trade policy outcome may be exactly the opposite – a reduction of trade barriers. See, for example, Gawande *et al.* (2004). But once again, the experience also points to a variety of experiences.

39 As Bhagwati pointed out, 'payments deficits and surpluses are macroeconomic phenomena that are not influenced in any predictable way by trade policy changes whose impact on the difference between domestic savings and invest-ment, if any, can come in different ways that can go in opposite directions' (cf. Bhagwati 1999, p. 11). This issue is discussed further below. For more details, see also Drabek (2004).

40 'Shallow' RIAs of course, also, incorporate an enforcement mechanism. It is argued here that these mechanisms cannot be as effective as in deep RIAs which are backed by strong political commitments.

41 This is precisely what has happened in most of the acceding Central and East European countries when joining the EU. The financial sector in these countries has been restructured with the help of all these policies.

42 These situations may include various kinds of market failure such as, for example, poor access to external borrowing or serious domestic structural rigidities. One such case is discussed, for example, by Begg in this volume.

43 The theoretical underpinning for the present argument is based on Kydland and Prescott (1977).

44 The reasons for failures of RIA, of course, vary from case to case but poor mac-roeconomic policies have typically been in the centre of problems. For more discussion see Genberg and de Simone (1993)

45 In her paper, she also reviews several theoretical arguments showing that har-monization of product standards can also have trade diverting features. However, the reviewed cases refer to situations with rather exceptional characteristics.

46 Hoekman (1998), p. 40.

47 For more details, see Bagwell and Staiger (1999), pp. 74-5.

48 Garoupa and Stephen have applied their argument to domestic law enforcement. However, their argument can be applied to international commitment whereby access to financial resources such *structural funds* in the EU can be seen as a proxy for a formal legal aid.

49 See discussion on p. 27 and de Melo, Panagariya and Rodrik (1993).

50 The relevant studies are reviewed in an older paper by Enders (1993). There are also few disadvantages of the dispute settlements mechanisms in regional settings which are reviewed by Enders.

51 This is, of course, a great simplification. NAFTA has gone much further than a simple free trade agreements as we shall see further below. For more details see also, for example, Hoekman (1998).

52 The classification made here is a variation of the 'model' proposed by Woolcock (2003).

53 See *Overview of Developments in the International Trading Systems*, Geneva: WTO, November 2002, WT/TPR/OV/8, p. 18.

54 See Spragia (2001) and Telo (2001a). Further deepening has been sought and proposed by President Fox. He suggested the establishment of a *Development Fund*, the full integration of wide-ranging labour and environmental issues into the agreement, support for less developed areas in the region, the expansion of coverage and competencies of NAFTA's 'weak' institutions, and the inclusion of new sectors and issues. See Chanona (2003), p. 8.

55 The preferential rules of origin agreed in NAFTA lead, according to some estimates, to costs that absorb up to 40 per cent of Mexico's preferential access to the US market. See Anson *et al.* (2003).

56 Most observers go, in fact, even further and argue that the AD measures have been trade diverting. See, for example, Carter and Gunning-Trant (2003).

57 See the discussion of the relevant point in the text above.

58 For more details see, for example, WTO (1996).

59 These are also countries that are covered in the empirical part of this volume.

60 See Feldman (2003) and his discussion of TBT issues.

61 Ibid.

62 Ibid.

63 The relationship between inflation and growth has been studied and the empirical evidence provided in a number of studies many of which are reported in Buiter, Lago and Stern (1997).

64 See the discussions in the following sections.

65 The theoretical conditions under which transition countries should be encouraged to impose (uniform) tariffs as a means of fiscal stabilization and increased competitiveness is discussed in this volume by Begg.

66 The experience with inflation targeting in Hungary, Poland and the Czech Republic is reviewed in Jonas and Mishkin (2003). They find that inflation targeting has been an effective pillar of the monetary strategy of these countries prior their accession to the European Union. For other relevant discussion see, for example, Nuti (2000).

67 However, a closer coordination of monetary policies does not eliminate all protectionist pressures. By tying their currency to the euro, countries still face the risk of exchange rate instability due to fluctuations in the relative values of the euro and the US dollar and other currencies. Nevertheless, the scope for currency instability is greatly reduced, given the trade exposure of the three countries with the European Union.

68 The Central and East European countries committed themselves under the Europe Agreements to adopt the EU rules concerning agreements between firms to restrict competition, abuse of dominant position, the behaviour of public undertakings and competition distorting state aids. A successful implementation of these legal instruments triggered off the commitment of the EU to discontinue recourse to anti-dumping.

69 Mavroidis and Neven (2000).

70 For more details see Michalek in this volume.
This point is also made by Fernandez and Portes (1998), p. 206. Together with other observers, Fernandez and Portes call these features 'non-traditional' gains and suggest that these gains may not be obtained unilaterally or multilaterally. Schiff and Winters (1998) make a similar point when they argue in favour of RIA on the grounds of increased security between nations. However, the original Europe Agreements did not contain all these WTO-Plus provisions. For example, the Central European countries' commitments on TRIPs, TRIMs and services in the Europe Agreements were derived from the relevant WTO agreements.

# References

Anderson, Kym and Richard Blackhurst, (eds) (1993) *Regional Integration and the Global Trading System*, New York and London: Harvester/Wheatsheaf.

Australia–USA Free Trade Agreement: Issues and Implications, (2001) A Report for the Department of Foreign Affairs and Trade, by the Australian APEC Study Centre, Monash University, August.

Anson, Jose, Olivier Cadot, Jaime de Melo, Antoni Estevadeordal, Akiko .Suwa-Eisenmann and Bolorma Tumurchudur (2003) 'Rules of Origin in North-South Preferential Trading Arrangements with an Application to NAFTA', London: CEPR Discussion Paper no. 4166, December.

Bagwell, Kyle and Robert W. Staiger (1999) 'Preferential Agreements and the Multilateral Trading System', in Baldwin *et al.* (1999), pp. 53–79.

Baldwin, Richard E. and Anthony J. Venables (1995) 'Regional Economic Integration', in Gene M. Grossman and Kenneth Rogoff (eds), *Handbook of International Economics*, New York and Oxford: Elsevier, 1995, vol. 3.

Bergloff, Erik and Ernst-Ludwig von Thadden (1999) 'The Changing Corporate Governance Paradigm: Implications for Transition and Developing Countries', Washington, DC: Annual World Bank Conference on Development Economics, Conference Paper, June.

Bhagwati, Jagdish (1999) 'Regionalism and Multilateralism: An Overview', in Bhagwati, Krishna and Panagariya (1999).

Bhagwati, Jagdish, Pravin Krishna, and Arvind Panagariya (1999) *Trading Blocks*, Cambridge, MA: MIT Press.

Bhagwati, Jagdish, and Arvind Panagariya (1999) 'Preferential Trading Areas and Multilateralism – Strangers, Friends, or Foes?', in Bhagwati, Krishna and Panagariya (1999).

Bussiere, Mathieu and Christian Mulder (1999) 'Political Instability and Economic Vulnerability', Washington, DC: International Monetary Fund, Policy Development and Review Department, Working Paper no. 99/46, April.

Buiter, Willem, Ricardo Lago, and Nicholas Stern (1997) 'Promoting Effective Market Economy in a Changing World', London: European Bank for Reconstruction and Development, Working Paper no. 23, April.

Carre, M., S. Levasseur and F. Portier (1996) 'Economic Integration, Asymmetries and the Desirability of a Monetary Union', Paris: *Economie Mathematique et Applications* Series no. 96.54.

Carter, Colin A. and Caroline Gunning-Trant (2003) 'Trade Remedy Laws and NAFTA Agricultural Trade', University of California at Davis: Working Paper no. 03-001, January.

Cecchini, P. *et al.* (1988): *The European Challenge 1992: Benefits of a Single Market*; Aldershot, Brookfield, USA, Hong Kong, Singapore and Sidney: Gower Publishing Company.

Chanona, Alejandro (2003) 'A Comparative Perspective between the European Union and NAFTA', Miami: University of Miami, mimeo, August.

Corden, Max (1994) *Macroeconomic Policy and Protection; Economic Policy, Exchange Rates and the International System*; Oxford: Oxford University Press, 1994.

Corsetti, Giancarlo and Paolo Pesenti (2002) 'Self-Validating Optimum Currency Areas', London: CEPR Discussion Paper no. 3220, 2002.

Crawford, Jo-Ann and Sam Laird (2000) 'Regional Trade Agreements and the WTO', Nottingham: University of Nottingham, CREDIT Research Paper no. 00/3.

Danthine, Jean-Pierre, Francesco Giavazzi and Ernst-Ludwig von Thadden (2000) 'European Financial Markets After EMU: A First Assessment', London: CEPR Discussion Paper no. 2413.

De Melo, Jaime, Arvind Panagariya and Dani Rodrik (1993) 'Regional Integration: An Analytical and Empirical Overview', in J. de Melo and A. Panagariya (eds), *New Dimensions in Regional Integration*, Cambridge: Cambridge University Press.

Dee, Philippa and Jyothi V. Gali (2003) 'Trade and Investment Effects of Preferential Trading Arrangements', Cambridge, MA: NBER Working Paper no. 10160, December.

Drabek, Zdenek (2004) 'International Trade and Macroeconomic Policy', Geneva: WTO, Economic Research Division, forthcoming Working Paper.

Drabek, Zdenek, and Josef C. Brada (1998) 'Exchange Rate Regimes and the Stability of Trade Policy in Transition Economies', *Journal of Comparative Economics*, vol. 26, pp. 642–68.

Drabek, Zdenek and Laird, Samuel (1998): The New Liberalism: Trade Policy Developments in Emerging Markets, *Journal of World Trade*, vol. 32, no. 5, (October)

Eliassen, Kjell A. and Catherine Borve Monsen (2001) 'Comparison of European and South East Asian Integration', in Telo (2001b).

Enders, Alice (1993) 'Dispute Settlement in Regional and Multilateral Trade Agreements', in Anderson and Blackhurst (1993).

Feldman, Magnus (2003) 'The Association Agreement between Poland and the EU', in Sampson and Woolcock (2003).

Feldstein, Martin (2000) 'The European Central Bank and the Euro: The First Year', Cambridge, MA: NBER Working Paper no. 7517, February.

Fernandez, Raquel and Jonathan Portes (1998) 'Returns to Regionalism: An Analysis of Non-Traditional Gains from Regional Trade Agreements', *World Economic Review*, vol. 12, no. 2, pp. 197–220.

Fisher, Stanley (1990) 'Rules versus Discretion in Monetary Policy', in B.M. Friedman and F.H. Hahn (eds), *Handbook in Monetary Policy*, Amsterdam: North-Holland, 1990.

Frankel, Jeffrey A. and David Romer (1996) 'Trade and Growth: An Empirical Investigation', Cambridge, MA: NBER Working Paper no. 5476.

Frankel, Jeffrey A., and Andrew K. Rose (2000) 'An Estimate of the Effects of Currency Unions on Trade and Growth', paper written for the Conference on Currency Unions, Stanford University, May 19–20.

Garay, Luis and Jorge Rafael Cornejo (2002) 'Rules of Origin and Trade Preferences', in Hoekman, B., A. Mattoo and P. English (eds) *Development, Trade and the WTO – A Handbook*, Washington, DC: World Bank, pp. 114–21.

Garoupa, Nuno and Frank Stephen (2003) 'A Note on Optimal Law Enforcement', London: CEPR Discussion Paper no. 4113, November.

Gawande, Kishore, Pravin Krishna and Michael J. Robbins (2004) 'Foreign Lobbies and US Trade Policy', Cambridge, MA: NBER Working Paper no. 10205.

Genberg, Hans and Francisco N. De Simone (1993) 'Regional Integration Agreements and Macroeconomic Discipline', in Anderson and Blackhurst (1993), pp. 199–217.

Geradin, Damien and Michel Kerf (2003) 'Levelling the Playing Field: Is the World Trade Organization Adequately Equipped to Prevent Anti-Competitive Practices in Telecommunication Practices?', Washington, DC: World Bank, Working Paper no. 2003 (Telecom).

Greenaway, David and Chris Milner (2002) 'Regionalism and Gravity', *Scottish Journal of Political Economy*, vol. 49, pp. 574–85.

Hallet, Andrew Hughes and Carlos A. Primo Braga (1994) 'The New Regionalism and the Threat of Protectionism', Washington, DC: World Bank, Policy Research Working Paper no. 1349.

Hare, Paul (2001) 'Trade Policy During the Transition: Lesson from the 1990s', *World Economy*, vol. 24, April, no. 4, pp. 483–511.

Hoekman, Bernard (1998) 'Free Trade and Deep Integration: Anti-Dumping and Anti-Trust in Regional Agreements', Washington, DC: World Bank, mimeo.

Hoekman, Bernard and Denise Eby Konan (1999) 'Deep Integration, Non-discrimination, and Euro-Mediterranean Free Trade', Washington, DC: World Bank, Policy Research Working Paper, no. 2130 May.

Hoekman, Bernard and Michael Leidy (1993) 'Holes and Loopholes in Integration Agreements: History and Prospects', in Anderson and Blackhurst (1993).

Imbs, Jean (1998) 'Co-Fluctuations', Lausanne, Switzerland: University of Lausanne, DEEP, HEC Working Paper no. 9819.

Jonas, Jiri and Frederic S. Mishkin (2003) 'Inflation Targeting in Transition Countries: Experience and Prospects', NBER, Working Paper no. W9667, May.

Kaminski, Bartolomej and Francis Ng (2001) 'Trade and Production Fragmentation: Central European Economies in European Union Networks of Production and Marketing', Washington, DC: World Bank Working Paper no. 2611, 2001.

Kenen, Peter (2002) 'Currency Unions and Trade: Variations on Themes by Rose and Persson', Princeton University and Reserve Bank of New Zealand Discussion Paper no. DP 2002/08.

Krueger, Anne (1993) 'Free Trade Agreements as Protectionist Devices: Rules of Origin', Cambridge, Mass., NBER Working Paper no. 4352.

Krugman, Paul (1991) 'Is Bilateralism Bad?', in E. Helpman, and A. Razin (eds), *International Trade and Trade Policy*, Cambridge, MA: MIT Press, 1991.

Krugman, Paul (1999) 'Regionalism versus Multilateralism: Analytical Notes', in Bhagwati *et al.* (1999), pp. 381–404.

Krugman, Paul and Anthony J. Venables (1990) 'Integration and Competitiveness of Peripheral Industry', in C. Bliss and J.C. Macedo (eds), *Unity with Diversity in the European Community*; Cambridge, UK: CEPR and Cambridge University Press.

Kydland, F.E. and E.C. Prescott (1977) 'Rules Rather Than Discretion: The Inconsistency of Optimal Plans', *Journal of Political Economy*, vol. 85, pp. 473–91.

Lawrence, Robert Z. (1996) *Regionalism, Multilateralism, and Deeper Integration*, Washington, DC: Brookings Institution.

MacBean, Alistair (2000) *Trade and Transition: Trade Promotion in Transition Economies*, London: Frank Cass.

Maskus, Keith and John Wilson (2001) *Quantifying the Impact of Technical Barriers to Trade: Can It Be Done?*, Ann Arbor: University of Michigan Press.

Mavroidis, Petros C. and Damien J. Neven (2000) 'The International Dimension of the Anti-Trust Practice in Poland and Hungary and the Czech Republic', Lausanne: Université de Lausanne, Ecole des HEC, Cahiers de Recherches Economiques du Department d'Econometrie et d'Economie Politique (DEEP) no. 00.16.

Messerlin, Patrick (2001) *Measuring the Costs of Protection in Europe: European Commercial Policy in the 2000s*, Washington, DC: Institute for International Economics.

Neven, Damien J. and Lars-Hendrik Röller (2000) 'The Scope of Conflict in International Merger Control', Berlin: Social Science Research Centre Discussion Paper no. FS IV 00–14.

Neven, Damien J. and Charles Wyplosz (1996) 'Relative Prices, Trade, and Restructuring of the European Industry', in *Trade and Jobs in Europe – Much Ado About Nothing?*, Oxford University Press, 1999.

Nuti, Mairo (2000) 'The Polish Zloty 1990–99: Success and Under-Performance', *American Economic Review*, May, papers and proceedings.

Oates, Wallace (1999) 'An Essay on Fiscal Federalism', *Journal of Economic Literature*, vol. 37, Issue 3, pp. 1120–49

Panagariya, Arvind (2002) 'On Necessarily Welfare-Enhancing Free Trade Areas', *Journal of International Economics*, vol. 57, August, no. 2, pp. 353–67.

Panagariya, Arvind and Rupa Dutta Gupta (2001) 'Free Trade Areas and Rules of Origins: Economics and Politics', Economics Working Paper Archive at WUSTL, International Trade no. 0308006.

Parsley, David C. and Hang-Jin Wei (2003) 'How Big and Heterogeneous are the Effects of Currency Arrangements on Market Integration?', Washington, DC: IMF, mimeo, (based on NBER Working Paper No. 8468).

Perotti, Enrico C. and Ernst-Ludwig von Thadden (2003) 'Strategic Transparency and Informed Trading: Will Capital Market Integration Force Convergence of Corporate Governance?', *Journal of Financial and Quantitative Analysis*, March, Special Issue.

Piermartini, Roberta (2004): 'Regionalism *a la Carte*', Geneva: WTO, mimeo, Research Department.

Portes, Richard and Helene Rey (1998) 'The Emergence of the Euro as the International Currency', Cambridge, MA: NBER Working Paper no. W6424.

Ritschl, Albrecht and Nicholas Wolf (2003) 'Endogeneity of Currency Areas and Trade Blocks: Evidence from the Inter-War period', London: CEPR Discussion Paper no. 4112, November.

Sampson, Gary and Stephen Woolcock (eds) (2003) *Regionalism, Multilateralism and Economic Integration: The Recent Experience*, Tokyo and Paris: United Nations University Press.

Sapsford, David, David Griffiths, and V.V. Balasubramanyan (2002) 'Regional Integration Agreements and Foreign Investment: Theory and Preliminary Evidence', *Manchester School*, vol. 70, pp. 460–82.

Schiff, Maurice (1996) *Small is Beautiful: Preferential Trade Agreements. Impact of Country Size, Market Share, Efficiency*, Washington, DC: World Bank, Working Paper No. 1668.

Schiff, Maurice and Alan Winters (1998) 'Regional Integration as Diplomacy', *World Bank Economic Review*, vol. 12, no. 2, pp. 271–96.

Scollay, Robert and John Gilbert (2001) *New Regional Trading Arrangements in the Asia-Pacific?*, Washington, DC: Institute for International Economics.

Sorsa, Pirita (1994) 'Regional Integration and the Baltics: Which Way?', Washington, DC: World Bank, Policy Research Working Paper no. 1390.

Spragia, Alberta M. (2001) 'European Union and NAFTA', in Telo (2001b), ch. 5.

Summers, Laurence (1999) 'Regionalism and the World Trading System', in Bhagwati *et al.* (1999).

Telo, Alberto (2001a) 'Between Trade Regionalization and Deep Integration', in Telo (2001b), pp. 71–95.

Telo, Alberto (2001b) *European Union and New Regionalism*, Aldershot: Ashgate.

Venables, Anthony J. (2003) 'Winners and Losers from Regional Integration Agreements', *Economic Journal*, vol. 113, October, pp. 747–61.

Viner (1950) *The Customs Union Issue*, New York: Carnegie Endowment for International Peace.

Whalley, John and J. Clark Leith (2003) 'Competitive Liberalization and a US-SACU FTA', Cambridge, MA: NBER Working Paper no. 10168, December.

Winters, L. Alan (1993) 'Expanding EC Membership and Association Accords: Recent Experience and Future Prospects', in Anderson and Blackhurst (1993).

Winters L. Alan (1999) 'Regionalism vs. Multilateralism', in Baldwin *et al.* (1999), pp. 7–49.

Woolcock, Stephen (2003) 'A Framework for Assessing RTAs: WTO-Plus', in Sampson and Woolcock (2003).

WTO (1995) *Regionalism and the World Trading System*, Geneva: WTO.

WTO (1996) *Annual Report*, Geneva: WTO.

WTO (2002a) *Overview of Developments in the International Trading Systems*, Geneva: WTO, November, WT/TPR/OV/8, p. 18.

WTO (2002b) *The European Union – Trade Policy Review 2002*, Geneva.

Wyplosz, Charles (2003) 'The Challenges of Wider and Deeper Europe', Vienna: paper presented to East-West Conference 2003, cosponsored by the Joint Vienna Institute and the Austrian National Bank, November 2–4.

# 2
# North, South, East, West: What's Best? Modern RTAs and their Implications for the Stability of Trade Policy

*Lucian Cernat and Sam Laird**

## Introduction

In the late 1980s and early 1990s, in parallel to the GATT negotiations under the Uruguay Round, many countries entered into trade negotiations aimed at the formation, revitalization or extension of regional trade agreements (RTAs). Some developed countries have consolidated their existing regional integration mechanisms going well beyond 'shallow integration', essentially limited to mutual tariff cuts. The EU Single Market of 1992 is a prime example of the new 'deeper integration'. Some countries, notably in Latin America, have revised moribund or ineffectual RTAs. Other countries in all regions have created new RTAs, or are currently involved in RTA formation. Most recently, RTAs have been initiated by countries that had traditionally been the main proponents of the multilateral approach under GATT (Japan, South Korea, Singapore and other countries in East Asia).

The integration process has progressed rapidly in many regions, especially in Europe and the Western Hemisphere. Regionalism, defined as both an increase in intra-regional trade flows and the number of regional trade agreements (RTAs) has intensified since the early 1990s (Figure 2.1). Currently the number of RTAs exceeds the number of WTO Members – many being party to a number of different RTAs in what Bhagwati calls a 'spaghetti bowl' of trade relationships (*The Economist*, March 3, 2001) – and the trend towards increased regionalism appears to be continuing. Among the early manifestations of this 'new regionalism' were NAFTA, MERCOSUR, the 'deepening' of EU integration through the Single Market programme, the EMU and the 'widening' of integration towards the East, all of which took place in a relatively short period. The countries in transition in Central and Eastern Europe have also adopted an active approach to regional integration, not

* UNCTAD, Geneva. Laird is also Special Professor of International Economics at the University of Nottingham. The views expressed are those of the authors and do not necessarily represent the views of the organizations to which they are affiliated.

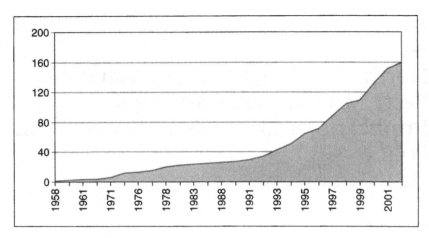

*Figure 2.1*   Number of existing RTAs
*Source*:   WTO.

only vis-à-vis the EU but also among themselves, reformulating their mutual relationships as market economies (cf. the old COMECON model).

Furthermore, the integration process has moved beyond the regional level to become inter-regional. New inter-continental integration projects with potentially significant impact on global trade and investment have been proliferating. APEC economies have agreed to achieve free and open trade and investment by 2010 (2020 in the case of developing countries) (UNES-CAP 1998). In the Western Hemisphere, the Free Trade Area of the Americas (FTAA), comprising 34 countries 'from Alaska to Tierra del Fuego', is in the making, with negotiations to be completed no later than 2005 (Aninat 1996; Devlin, Estevadeordal and Garay 1999). EU-induced regionalism has extended to countries and regions outside of Europe. The EU has plans and in some cases has already concluded free trade agreements with certain Commonwealth of Independent States (CIS) countries, with Mediterranean countries (including the EU – Turkey Customs Union), MERCOSUR, South Africa and Mexico. Outside the EU framework, other prominent cases of inter-continental RTAs involve various Latin American and Asian countries.

In parallel, regional integration agreements among developing countries have expanded, increased and in general gained new momentum. Impetus was to some extent provided by the dramatic liberalization of import regimes in developing countries consequent to structural adjustment programmes (Drabek and Laird 1998). In Latin America, MERCOSUR and the Andean Community have moved rapidly ahead with the implementation of their programmes to liberalize mutual trade and establish custom unions. CARICOM has been revitalized. In Asia, ASEAN has accelerated the implementation of its free trade area in goods and started work on liberalizing trade in services. In the Pacific, several countries have formed a free trade area within the Melanesian Spearhead Group and the Pacific Forum has agreed to form a free trade agreement. In Africa, several groupings have been engaged in major revisions and restructuring of integration such as UDEAC into CEMAC, and others are intensifying sub-regional integration such as SADC's adoption of its trade protocol in 2000 (calling for the formation of an FTA within eight years), and the entry into force of the COMESA FTA in 2000.

Mixed RTAs (North–South RTAs) with reciprocal commitments between developed and developing countries are becoming more frequent in all regions. These include, notably, agreements with the United States and the EU as the 'Northern' partner, but also include agreement with Australia, Canada, New Zealand and South Africa with developing countries in their regions.

Overall, RTAs are increasingly expanding to other regions and becoming more complex interregional integration systems with various grades and types of association (Figure 2.2).

The new dynamism in RTAs points to certain emerging results in the system of international trade relations. The creation and rapid expansion of RTAs seems set to remain a lasting feature of international economic relations. Some of the regional projects would combine substantial economic

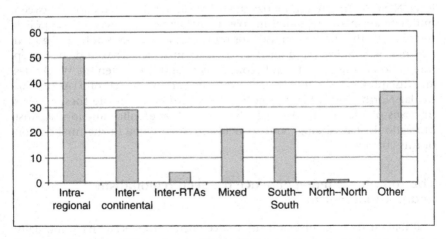

*Figure 2.2* Trends in RTA formation: RTAs at different stages of negotiations

*Source*: Cernat (2002).

*Legend*: **Intra-regional RTAs** refer to agreements between countries belonging to the same geographical region; **Inter-RTAs** refer to those trading arrangements between two or more existing RTAs; **Inter-continental RTAs** counts the number of new RTAs between countries situated on different continents; **Mixed RTAs** refer to agreements involving developing and developed countries, while **South–South** and **North–North** refer to developing-only and developed-only agreements, respectively. **Other** refers to RTAs involving transition economies.

power and would exert a major impact on third countries and on the functioning of the multilateral trading system. In terms of the international trading system, the concern is that the proliferation of RTAs may lead to an erosion and possible fragmentation of the multilateral trading system into some kind of federal system composed of semi-autonomous 'stumbling' trading blocs. This concern has lead to debate over regionalism versus multilateralism, and whether the former is a 'building block' or a 'stumbling block' towards the latter.

The evidence is mixed. RTA formation increases the interdependence between countries. The literature on economic interdependence points out that the effects of interdependence are sometimes contradictory: on one hand they enhance cooperation, but on the other hand they can increase tensions between partners. Transatlantic relations are probably the best example of interdependent economic actors engaged in both cooperative and conflictual relations.

This chapter seeks to clarify whether RTAs contribute to the smooth functioning of the multilateral trading system, in particular to what extent RTAs can prevent protectionist 'backslidings'. The chapter tries to identify several RTA-specific issues that are likely to enhance the functioning of the multilateral trading system, particularly in areas in which the WTO rules need to be strengthened. The remainder of this chapter is organised as follows. Next, we briefly summarize the debate about the interactions between multilateralism and regionalism. We then elaborate on several trade policy instruments that have been used for protectionist purposes at both regional and multilateral level and analyse the ways in which RTAs and their deep integration commitments could complement and strengthen the WTO rules and disciplines. Following this, we compare various models of RTAs (in particular North–North, South–South, and mixed) and evaluate these models in terms of the extent to which they act as 'trade policy anchors' against protectionist backsliding. Some conclusions are offered in the final section of this chapter.

## The complex interdependence between regionalism and multilateralism

Multilateral trading rules and rules developed within RTAs have mutually reacted during the last decades of their coexistence. On one hand, the multilateral rules have had a constraining effect (policy anchor) on regional arrangements, most obviously through the provisions of GATT Article XXIV, its post-Uruguay Round understanding, and GATS Article V. On the other hand, regional integration initiatives have also influenced the multilateral system in a number of ways, particularly in the establishment of rules in areas not yet covered by the WTO or in making clarifications about the operation of WTO rules. Thus, RTAs may act as *policy transfer mechanisms* towards the multilateral system. As a result of developments within RTAs, the WTO agenda has expanded, and new or more far-reaching rules have been introduced on trade-related issues along the lines of RTA provisions. The areas of investment and competition policy are areas where RTAs have moved ahead of the WTO system, while developments on services at the WTO level were influenced by progress in NAFTA and the EU.

As the subsequent analysis shows, the possibility of RTAs setting the pace on new rules or clarifications of existing rules is more likely on those issues where the WTO rules contain 'soft' provisions. Apparent 'backsliding' from multilateral commitments occurs for a number of reasons, including, importantly, the existence of loopholes in the regulatory framework. WTO provisions still provide a relatively wide degree of latitude for members on the use of certain export- and import-related measures. Examples of such latitude on export-related measures include agricultural export subsidies. Similarly, there is latitude for the application of state aids and investment measures to exports of industrial products, both of which are being regulated under the WTO agreements. Examples of import-related legalized 'backsliding' include anti-dumping, countervailing and safeguard provisions.

However, while the application of certain WTO rules – other than the MFN principle – may appear to be strengthened in practice, there is no *a priori* reason to expect this. For example, GATT Article XXIV requirements for customs unions and FTAs show no obligation to ensure that weak implementation of certain WTO provisions is better tackled under a regional agreement. Nor there is any obligation or recommendation that RTAs should be at the forefront of liberalization in those areas in which multilateral trade liberalization has failed so far to advance more rapidly. Yet, in some instances, a number of RTAs appear to fulfil (albeit sometimes partially) these requirements of being 'WTO – plus'.

This does not mean RTAs are not *stumbling blocs* for the multilateral trading system. Under RTAs, the MFN principle is weakened, although it has been observed that in some instances the arrangements seem to stimulate investment and economic growth, pulling in even higher trade from third countries than prior to the agreement (Crawford and Laird 2000). On the other hand, some arrangements, also intended to foster trade between partners – mutual recognition agreements (MRAs) on standards, for example – may operate to reinforce discrimination against third countries.

The WTO secretariat has conducted a number of studies aimed at assessing the extent to which regional trading arrangements have systemic implications on the multilateral trading rules. One such study (WT/REG/W/26) made an inventory of non-tariff provisions included in RTAs, in particular on quantitative restrictions, contingency instruments, rules on subsidies, technical barriers to trade, and standards. A previous study (WT/REG/W/8) looked in more detail at the existence and the actual wording of TBT provisions in a number of RTAs. However, both studies were descriptive and with few exceptions, did not attempt to gauge whether these provisions are 'WTO – plus', mere duplication, or less ambitious than existing WTO rules.

To look more closely at the impact of regional trading arrangements on the functioning of the multilateral trading system, the following section will look at specific sectoral arrangements under particular RTAs that may be more or less favourable to WTO objectives and rules, and whether the RTA provisions can act as policy-stabilizing mechanisms at multilateral level. The discussion will be centred on a number of specific RTAs, whose provisions make them relevant for the purpose of this analysis.

## Deep regional integration and multilateral trade rules: specific trade instruments

In recent years, both multilateralism and regionalism evolved to steps towards integration that go beyond tariffs or non-tariff border measures. Deep integration, defined as 'beyond-the-border measures' (Lawrence 1996; Mikesell 1963), is becoming an essential feature of both globalization and regionalization, and has tempered our understanding of regionalism: Regionalism has moved far beyond pure trade/tariff or market integration in the form of free trade areas or customs unions. Integration has now become much deeper, much more multifaceted and multisectoral, encompassing a wide range of economic and other political objectives (Bora and Findlay 1996; Whalley 1996). New RTAs place considerable emphasis on liberalization of services, investments and labour markets, government procurement, strengthening of technological and scientific cooperation, environment, common competition policies or monetary and financial integration. These are among the distinguishing components of NAFTA, FTAA, APEC, EU and its partnership agreements, and several agreements among developing countries. Yet, whether the formation of an ambitious RTA covering all these issues will reduce the chances of trade policy backsliding at multilateral level is still unclear. Both the multilateral trading rules and regional agreements still contain a number of loopholes that may lead to the re-emergence of protectionist actions against RTA partners or third countries.

Apart from a number of obvious trade policy instruments that have been used for protectionist purposes both at multilateral and regional level (such as anti-dumping, safeguard measures, countervailing duties, technical barriers to trade, customs procedures and rules of origin), the subsequent discussion will also cover a number of trade-related issues such as investment, competition policy and labour mobility. The reason for including these issues in a discussion about regionalism and the potential for protectionist backsliding at multilateral level is that protectionism may occur not only through trade policy backsliding *per se*, but also by enacting trade-related policies that restrict the movement of factors of production (capital and labour) or allowing private business anti-competitive practices that may impair the benefits expected from trade liberalization.

Lastly, dispute settlement procedures also deserve special attention. For both traditional trade policy and trade-related policy instruments, policy backsliding can be avoided or deterred when effective dispute settlement mechanisms are in place, either at multilateral or regional level.

The subsequent sections look at the experience of various RTAs with deeper integration measures with a view to determining the extent to which such RTAs have an influence on multilateral trade disciplines and whether these influences are positive or negative.

## Anti-dumping

The elimination of anti-dumping measures in RTAs is an exception rather than a rule.[2] So far, only three North–North RTAs (EU, EEA and ANZCERTA) and two North–South regional agreements (Canada–Chile and Canada–Costa Rica FTA) have eliminated anti-dumping among participants. In the case of MERCOSUR, the parties have discussed the idea of anti-dumping but economic difficulties have confronted the advancement of the agreement.

In the case of developing countries, several South–South RTAs have explicitly provided for the applicability of anti-dumping on trade among members. In the case of CARICOM, for instance, WTO rules have been explicitly included in the protocol signed in March 2000, which amends the original CARICOM treaty on matters related to competition policy, consumer protection, dumping and subsidies.[3]

Most regional agreements allow the use of anti-dumping measures among their members according to WTO rules. Anti-dumping provisions are preserved and used more often in North–South RTAs. In NAFTA, for instance, anti-dumping and countervailing duties are still applicable, and have been used on mutual trade.

However, one particular NAFTA provision appears to provide for the exemption of Canada and Mexico from anti-dumping measures, countervailing duties or even safeguard measures by requiring that any NAFTA measure *specifically* names other NAFTA parties before it applies to them (Kerr, 2001:1174). In other words, unless explicitly mentioned, both Canada and Mexico are by default excluded from such trade measures. (The specificity of anti-dumping measures – relating to individual countries or even firms – is quite different from safeguards that are general in nature.)

The Europe Agreements also maintain both anti-dumping and safeguard measures as policy options between partners. Furthermore, other than in the cases where anti-dumping is specifically precluded, there is little evidence that RTAs contribute towards the elimination of anti-dumping cases among RTA members (Hoekman 1998). In many cases where anti-dumping actions between RTA members are left in place, anti-dumping measures against

*Table 2.1* Anti-dumping actions among certain RTA members

| Regional agreement | Countries | | Number of anti-dumping actions | |
|---|---|---|---|---|
| | Initiating | Against | 07.97–07.98 RTA member/ Total | 07.98–07.99 RTA member/ Total |
| EU Agreements | EU | EU Associate countries | 10/44 | 9/41 |
| | Czech Republic | EU countries | – | 2/2 |
| | Poland | Germany | – | 1/2 |
| Mercosur | Argentina | Brazil | 3/8 | 5/15 |
| NAFTA | USA | Mexico | | 2/43 |
| | Mexico | USA | 1/8 | 6/12 |
| | USA | Canada | 2/28 | – |
| | Canada | USA | 1/10 | – |

*Source*: WTO.

RTA members account for a large share of the total anti-dumping measures adopted (Table 2.1).

### Subsidies and countervailing duties

Several RTAs have gone beyond WTO disciplines in the area of subsidies. In both EU and the EEA, state aids affecting trade flows are prohibited between members, although general available subsidies are permitted in principle. A similar approach has also been adopted by CEFTA, where all subsidies affecting trade flows have been eliminated in internal trade. Nevertheless, in the case of both the EEA and CEFTA, subsidies can be used according to WTO rules for agricultural products. ANZCERTA also includes disciplines on subsidies (Article 11) that are stronger than those contained in the WTO. All export subsidies were eliminated in internal trade by 1987 (WTO 2002c:5). Regarding domestic support, following the first five-year General Review of ANZCERTA in 1988, the two participants signed an Agreed Minute on Industry Assistance under which they agreed not to pay (from July 1990) production bounties or similar measures on goods exported to the other member and undertook to attempt to avoid the adoption of industry-specific measures (bounties, subsidies and other financial support) that have adverse effects on competition within the FTA.

Similarly to ANZCERTA, the Canada–Chile FTA eliminates the possibility of using export subsidies in the agricultural sector. Article C-14 of the Agreement stipulates that no member is allowed to maintain export subsidies in internal trade after January 1, 2003. This is also true for the Canada–Costa Rica FTA where export subsidies for agricultural goods have been eliminated since the entry into force of the agreement.

## Safeguard measures

One rationale among developing countries in particular when entering into mixed agreements with their main developed trading partners is to open up these markets for sensitive products by removing tariff peaks or non-tariff barriers. The value of this improved market access is nevertheless reduced by other measures that remain in place among RTA members, especially safeguard measures.

The rules of the World Trade Organization recognize that sometimes imports, whether fairly or unfairly traded, can cause such harm to domestic industries that temporary restraints are warranted. And these rules include safeguard provisions for industries that have had substantial injury from imports. Some countries – for example, Japan, Korea, India, the United States, European Union, Brazil – have repeatedly used safeguards in recent years.

RTAs have dealt with safeguards in a variety of ways. Some RTAs apply the WTO or less stringent rules (CEFTA, for instance), others have strengthened their applications (NAFTA, the EU-Mexico FTA),[4] while few RTAs have abolished safeguards altogether on trade between members. Similarly to the EU on its internal trade, other agreements such as ANZCERTA, the agreement between New Zealand and Singapore on a Closer Economic Partnership (CEP), and MERCOSUR have eliminated safeguard measures.

There is some disagreement as to whether, under the WTO rules, RTAs should be allowed to apply safeguards, subject to certain conditions, on non-members only. Article 2, paragraph 2, of the Agreement on Safeguards states that safeguards should be applied to imports irrespective of their source. However, a footnote to paragraph 1 of the same article stipulates that 'nothing in this Agreement prejudges the interpretation of the relationship between Article XIX and paragraph 8 of Article XXIV of GATT 1994'. This issue arose in the *Argentina Footwear* case.[5] The crucial issue raised was exactly the relationship between Article XXIV, on one hand, and Article XIX of GATT and Article 2.2 of the Agreement on Safeguards, on the other.[6] Although the safeguards measures applied by Argentina on non-members only were found to violate the WTO rules, the Appellate Body indicated that no ruling was made on 'whether, as a general principle, a member of a customs union can exclude other members of that customs union from the application of a safeguard measure'.

A recent case involving discriminatory safeguards is the recent case of US safeguards on steel imports. The USA announced that it had decided to exclude its RTA partners (Canada, Mexico, Israel and Jordan)[7] and certain developing countries[8] from the steel safeguard measures.[9] Other regional arrangements do not exclude partners from global safeguard measures. Under the EEA, for instance, despite the high level of integration and unlike the USA under CUSFTA, the EU did not exclude the EFTA States from the safeguard measures intended to countervail the potential surge in imports as a result of US trade actions.[10]

Under the Canada–Chile Free Trade Agreement (CCFTA) bilateral imports are exempted from safeguard measures unless these imports contribute importantly to the serious injury of the domestic industry. The CCFTA establishes that when an exemption is not granted, the 'Party taking action . . . shall provide to the other Party mutually agreed trade liberalizing compensation in the form of concessions having substantially equivalent trade effects' (Art. F-02.6).[11]

### Technical barriers to trade (TBTs) and standards

The WTO Agreement on Technical Barriers encourages members to harmonize their technical regulations and use international standards in their trade. However, these provisions are watered down by several escape clauses that can be used in a discretionary way. One way in which several RTAs, both North–North and South–South types, went beyond the WTO rules was by providing for harmonization of standards and mutual recognition of national standards among RTA members. This is the case of the EU, EEA and EFTA. One important element in the reduction of the negative impact of TBTs on trade flows consists of mutual recognition of conformity assessment. Such an agreement is sometimes concluded even without the framework of an RTA, as in the case of the agreement between the USA and EU.

In other cases, even advanced RTAs maintain technical barriers and cumbersome standards that may act as a barrier to intra-regional trade flows, despite enhanced mechanisms for cooperation at regional level. Kerr (1997) brings detailed evidence that, in the case of NAFTA, regional integration is not necessarily a more efficient strategy to eliminate trade-distorting TBTs and health, sanitary and phytosanitary standards on agricultural products.

### Trade facilitation

RTAs may provide for mutual recognition of formalities carried out by the competent authorities of the other parties. Several African RTAs have been instrumental in introducing new trade facilitation measures among members. For instance, ECOWAS introduced harmonized customs documents, a region-wide vehicle insurance scheme.[12] Similarly, SADC introduced several customs and trade facilitation initiatives at regional level, such as the issuance of harmonized SADC customs documents. In addition, SADC members have agreed to eliminate cumbersome import and export licensing and permits, unnecessary import and export quotas, import bans and prohibitions. Efforts are being made to eliminate visa requirements and other custom-related trade barriers.[13]

The EFTA agreement provides for mutual recognition of inspections carried out and of documents certifying compliance with the requirements of the import country or equivalent requirements of the export country. ASEAN countries have concluded an agreement for the recognition of commercial vehicle inspection certificates for goods vehicles used for transit transport. MERCOSUR has established a series of agreements ensuring cooperation between customs authorities, including the 1993 Recife agreement for coordinating border controls, which establishes technical and operational measures to regulate the functioning of integrated border controls.[14]

### Trade-related investment measures

The adoption of the WTO Agreement on Trade-Related Investment Measures brought several new disciplines in the multilateral trading system. In general, the new WTO rules on trade-related investment measures had been already implemented in a number of North–North RTAs. For instance, the provisions of the NAFTA concerning performance requirements apply to both investments of investors from NAFTA members and investors from third countries. However, even in the case of North–North RTAs, several exceptions to the WTO disciplines were still permitted.[15]

The TRIMs Agreement had a greater impact on developing countries, which for long maintained various forms of regulatory investment policies aimed (with more or less success) at fostering industrialization (Bora, Lloyd, and Pangestu, 2000). Such measures were particularly important in the automotive sector. Despite these new multilateral constraints on the policy options available for developing countries to regulate foreign investment, several regional agreements aim at fostering cooperation between members by establishing a special legal regime for the formation of a regional form of business enterprise.

For example, the Uniform Code on Andean Multinational Enterprises established by Decision 292 of the Commission of the Cartagena Agreement provides for the formation of Andean Multinational Enterprises.[16] One of the conditions for the creation of such an enterprise is that capital contributions by national investors of two or more member countries must make up more than 60 per cent of the capital of the enterprise. Among the privileges which the Decision requires member countries to grant to such enterprises are national treatment with respect to government procurement, export incentives and taxation, the right to participate in economic sectors reserved for national companies, the right to open branches in any member country, and the right of free transfer of funds related to investments. Likewise, the Basic Agreement on the ASEAN Industrial Cooperation Scheme (AICO Scheme) was concluded by members of ASEAN in 1996 to promote joint manufacturing industrial activities between ASEAN-based companies.

Several African regional initiatives also contain provisions aimed at promoting intra-regional investment. The Treaty Establishing the African Economic Community (1991) and the Treaty Establishing the Common Market for Eastern and Southern Africa (COMESA) (1993) include among their objectives the removal of obstacles to the free movement of capital and the right of residence and establishment. Finally, the Revised Treaty of the Economic Community of West African States (ECOWAS) (1993) includes among its objectives the establishment of a common market involving, inter alia, the removal of obstacles to the free movement of persons, goods, services and capital and obstacles to the right of residence and establishment (Article 3(2)). The Treaty Establishing the Economic and Monetary Union of West Africa (1996) provides for freedom to provide services in the territory of another member State and proscribes restrictions on movement of capital.

The lack of detailed information on the implementation of these South–South initiatives makes it difficult to estimate the actual impact of such ambitious provisions on investment flows among RTA members until reliable data become available.

### Trade and competition policy

In contrast to dumping and subsidies, which are covered by multilateral WTO rules, no attempt has been made – since the abandonment of Chapter V of the Havana Charter on restrictive business practices – to introduce general commitments on competition policy at multilateral level, although there are competition policy aspects of the TRIPS Agreement and the GATS in particular.

However, several RTAs have gone beyond the WTO rules in promoting more far-reaching rules on trade and competition policy. Prime examples are the European Union and ANZCERTA who adopted two different integration approaches. In the European Union, the European Commission is the supranational authority that ensures that competition policy is enforced throughout the Union. The EEA also extended the EU competition policy to EFTA countries, with the European Commission and the European Surveillance Authority sharing the enforcement responsibilities on competition-related issues. As the EU Treaties, the EEA prohibits price fixing, abuse of dominant position, or any other practices that may affect trade between parties.[17]

Under ANZCERTA, each country's competition authority and courts have a unique model of 'overlapping jurisdiction' – whereby each competition authority may control the misuse of market power in the trans-Tasman market. The agreement provides for extensive investigatory assistance, the exchange of information (subject to rules of confidentiality) and coordinated enforcement. The experience of ANZCERTA is quite illustrative of the way an RTA can avoid the use or misuse of anti-dumping practices, a common 'backsliding' problem at multilateral level. Since the adoption of ANZCERTA, not a single case of trans-Tasman anti-competitive practices has been investigated by the competition authorities vested with their new regional enforcing powers. This stands in sharp contrast with the active use of anti-dumping investigations between Australia and New Zealand prior to ANZCERTA. The EEA and ANZCERTA have in common the elimination of anti-dumping rules and their replacement with regional rules on competition policy. The EEA is the only RTA concluded by the EU where anti-dumping and countervailing duties are eliminated and replaced by competition rules.

Not only North–North RTAs but also several North–South RTAs agreements are advancing the agenda on trade and competition policy. Two RTAs concluded by Canada (Canada–Chile and Canada–Costa Rica FTA) provide for a concrete framework for cooperation and consultation and enhancement of the effectiveness of enforcement activities by competition authorities. Although less ambitious than ANZCERTA and EEA, the Europe Agreements between the EU and candidate countries also contain provisions regarding anti-competitive practices. However, unlike ANZCERTA, the Europe Agreements state that candidate countries should harmonize their competition laws with those of the EU, and that each national competition authority will settle anti-competitive cases in accordance with national rules. For most of these countries, the Europe Agreements were a major incentive to adopt domestic competition laws for the first time and create appropriate institutional infrastructures. The FTA between Japan and Singapore, which calls for coordinated enforcement of competition policies, provides a similar example.

Among South–South RTAs, only a few initiatives have a regional institutional framework to deal with competition policies. This is to a large extent due to the fact that, in many developing countries, competition laws and authorities are non-existent or underdeveloped. The Andean Community provides one notable exception among South–South RTAs. Similar to the EU approach, the Andean Community institutions also have supranational powers, as the Board of the Cartegena Agreement is assigned the responsibility to investigate alleged anti-competitive infringements, and its subsequent orders have direct legal effect in member countries. Competition law and policy is starting to be addressed more extensively in African sub-regional agreements. The Treaty Establishing the Common Market for Eastern and Southern Africa (COMESA), for instance, prohibits in Article 55 'any agreement between undertakings or concerted practices which has as its objective or effect the prevention, restriction or distortion of competition within the Common Market'. Furthermore, COMESA contemplates formulating and implementing a regional competition policy which will harmonize existing national competition policies, or introduce them where they were absent, in the context of a transition to a full customs union (Musonda 2000). MERCOSUR also aims to introduce a regional approach to competition policy.

## Trade in services and labour mobility

Several RTAs involving both developed and developing countries contain provisions not only on trade in goods, but also on trade in services.[18] NAFTA, MERCOSUR, SADC, CARICOM, EU – Mexico FTA, all contain rules, disciplines and liberalization commitments on trade in services at regional level. Other regional integration arrangements, such as SAARC, have made little progress to include services trade on their agenda.[19]

In principle, WTO members of a regional agreement in services should benefit at least from MFN treatment plus whatever is agreed regionally and notified as an MFN exemption. However, the extent to which the access offered by individual parties goes beyond their GATS commitments would need a case-by-case analysis. The GATS and most regional initiatives have taken a positive list and sectoral approach, nominating areas where commitments are made, while NAFTA has taken a negative list approach under which all areas are covered unless explicitly excluded. While proponents argue that the negative list approach encourages participants to take on greater commitments, there is no hard evidence of this. In practice, in the WTO and regional agreements, the initial coverage of services was largely confined to existing practice, and it will take time to assess which approach leads to a faster widening and deepening of commitments.

A key issue affecting trade across many services sectors is labour mobility. The experience of South–South RTAs with labour mobility is mixed. Some RTAs, like MERCOSUR, for instance, follow the GATS model closely. Other RTAs have more ambitious goals. COMESA, for example, has 'full labour mobility' as the agreed objective of the agreement (see Article 164 of the COMESA Treaty), although progress towards that objective appears to have been limited to date. The ultimate aim of SADC is to promote the free movement of goods and services within the region. However, there are currently no provisions for free movement of labour or service suppliers. ECOWAS's Market Integration Programme has achieved significant progress, among others, in the areas of free movement of persons, abolition of visas and entry permits, and introduction of harmonized immigration and emigration forms.

In 1989, CARICOM agreed to eliminate work permits for CARICOM nationals. Labour mobility was further enhanced after the adoption of Protocol II in 1998. The agreement provides for free movement of university graduates, other professionals and skilled persons, and selected occupations as well as freedom of travel and exercise of a profession (i.e. elimination of passport requirements, facilitation of entry at immigration points, elimination of work permit requirements for CARICOM nationals). Furthermore, specific provisions require CARICOM members to ensure mutual recognition of diplomas, certificates and qualifications.

A similar mixed experience is found among North–North RTAs. The EU and EEA provide for full labour mobility. On the other hand, ANZCERTA does not cover general labour mobility but this is not needed, as, under the 'Trans-Tasman Travel Arrangement', Australians and New Zealanders are free to live and work in each other's countries for an indefinite period. As with other North–South arrangements, NAFTA provides for a series of labour movement facilities. Chapter 16 of NAFTA facilitates movement of business persons. Access is basically limited to four higher skills categories: traders and investors, intra-company transferees, business visitors and professionals.

The Canada–Chile Free Trade Agreement adopts a similar approach. While some agreements (e.g. EU) allow for general mobility of people and confer immigration rights, the majority of agreements provide only special access or facilitation of existing access within existing immigration arrangements. In most agreements, labour mobility does not override general migration legislation and parties retain broad discretion to grant, refuse and administer residence permits and visas.

Such variety of labour movement provisions may be explained by a number of factors: economic disparity, geographical proximity and domestic labour market conditions. In those RTAs where labour movements are strong, market access for developing countries' workers are more difficult, to the extent that developed countries maintain overall ceilings on access of foreign labour and immigration that are reserved primarily for nationals from member states of the RTA. Common rules and procedures for immigration within a large RTA may further tend to reduce access by developing countries' labour, if visa and immigration controls are extended to a larger number of third countries and applied also by hitherto more liberal member states. On the other hand, tightening the application of labour standards within large RTAs may reduce international competitiveness of a member country.

### Dispute settlements

RTAs have adopted two broad strategies to address bilateral trade disputes. One strategy is a legalistic, formal dispute settlement process, either in the form of a judicial body (the European Court of Justice, in the case of the EU) or through arbitration panels, as in the case of NAFTA. NAFTA includes at least five distinct mechanisms for different issue areas: general disputes (Chapter 20), unfair trade laws (Chapter 19), investment (Chapter 11), and the side accords on labour and the environment. However, many RTAs do not have a legalistic dispute settlement mechanism. Instead, they rely on a more diplomatic mechanism. Trade disputes are referred to a joint body (often called the Joint Committee) to solve trade disputes between parties.

It is difficult to say which model is more appropriate to diffuse trade disputes. Information on disputes referred to joint committees is not readily available and therefore their efficiency or deterrent properties cannot be assessed. On the other hand, some RTAs that rely on joint committees rather than on formal legalistic procedures (such as the Europe Agreements for instance) have so far avoided trade disputes at multilateral level. Unlike the Europe Agreements, other RTAs have not excluded the potential for acute trade tensions among their members, neither at multilateral nor at regional level. Probably the most prominent example is NAFTA where a number of disputes, like recent cases on sugar, trucking and US extra-territorial rules, could not be papered over by NAFTA members. Under NAFTA, Mexican trucks were supposed to be allowed to travel throughout the United States by

1 January, 2000, but union and safety groups have kept that from happening. Both countries have been arguing about a series of sweetener tariffs since Mexico imposed anti-dumping duties on corn syrup imports from the United States in 1997. The anti-dumping duties were not lifted until two WTO panels found Mexico's anti-dumping import duties on high fructose corn syrup from the United States in breach of multilateral rules. The tariffs came in retaliation for the United States's limit on the amount of tariff-free sugar imports from Mexico.

Another tension between NAFTA partners was induced by the Helms – Burton Act that, under the US extra-territorial law doctrine, threatens with sanctions such as possible refusal of visas and the exclusion of non-US nationals from US territory if they breach the unilateral trade restrictions imposed by the USA on Cuba. Several WTO members argued that the Helms–Burton act infringes WTO rules. Similarly, both Canada and Mexico argue that the Act goes against several NAFTA provisions by threatening to deny NAFTA businessmen entry to the United States and by breaking investor-protection guarantees.

### Rules of origin (ROOs)

The operation of rules of origin is one of the areas of greatest concern in RTAs, having considerable scope for trade diversion. Even today, the WTO has no provisions on the use of either preferential or non-preferential ROOs, and discussions continue to be blocked in this area. Accordingly, WTO members are free to use a variety of methods for the determination of origin.[20] Some South–South RTAs, as in the case of COMESA, have adopted more liberal and simple rules of origin than North–North RTAs.[21] In many RTAs, rules of origin have become captive to special interest lobbies and are used as protectionist devices. Stringent rules of origin can have a similar effect on trade as high tariffs, if the effect is precluding a producer from using the most efficient source for their inputs. For example, the rules of origin for apparel under NAFTA essentially forbid the use of imported fabrics, yarns and even some fibres in the manufacture of qualifying apparel.

The increased importance of rules of origin has determined certain producers to make use of specialized firms that give tailor-made advice on the right mix of inputs that qualify final products for the preferential regime.[22]

One notable exception to this tendency of complex rules of origin is provided by those rules of origin that provide for cumulation of origin between RTA partners and third countries. Some GSP schemes allow for cumulation of origin between beneficiary countries, or between LDCs and non-LDCs that are part of regional integration initiatives. A large number of EU-generated RTAs have become part of the pan-European system of cumulation of origin, which essentially creates a wide free trade area with harmonized rules of origin.

ROOs are therefore an area where RTAs are again outpacing the WTO system, but it is difficult to argue that, as they are used, they are necessarily an advance on the system. Indeed, the diversity of practices suggests an urgent need for a multilateral agreement in this area.

## RTAs as 'building blocks': North–North, South–South or mixed?

The evidence surveyed above suggests that the approach taken by RTAs to trade rules is quite varied. Some RTAs have made clear steps towards trade liberalization beyond existing WTO rules.

The WTO examination process sheds some insights into the operation of a number of agreements, principally where a developed country is involved; that is, North–North, North–South and North–East (agreement between developed countries and economies in transition). This is because agreement between developing countries under the Enabling Clause are not subject to an examination process. This process and other published studies show that the application of deep integration provisions are most advanced in agreements involving developed countries, which are pushing their partners in these areas, although sometimes with longer transition periods (asymmetry). Such schemes are therefore the most important driving forces pushing forward the agenda at the multilateral level.

In contrast, as many authors have suggested, most RTAs among developing countries are still in the realm of shallow integration (understood as removal or reduction of border measures), with little progress towards deeper integration, even where this is envisaged in the agreements; for example, in MERCOSUR. Even when the latter becomes a priority, often incomplete shallow integration limits the prospect of deeper integration. Consequently, it has been argued that the developing countries have not reaped the full potential advantages from integration in terms of export diversification, increased international competitiveness, more efficient allocation of resources, or significant stimulation of production and investment in the region (Foroutan 1993; Nogues and Quintanilla 1993; Yeats 1998). On the one hand, the lesser (more cautious) degree of implementation among developing countries is related to their economic situation, sometimes with large fluctuations in trade flows resulting from exchange rate movements and relative macroeconomic instability. On the other hand, their stage of development suggests the need for some flexibility or policy space in their trade and sectoral policies as adjustments to greater openness – whether in a regional context or more generally – often have initial negative consequences.

It has been suggested that integration would be fostered by greater use of common currencies (Rose 2002). This could certainly reduce transactions costs for trade, irrespective of the formal nature of the trade partnership. However, substantial macro-economic stability and convergence vis-à-vis major trading partners are certainly priors to any adoption of a common currency. The targeting approach of the EU as a basis for the adoption of the euro is an example. The Asian crisis is also an example where, when economic fundamentals start to vary widely, locking into a dollar anchor – which is analogous although not identical to a common currency for trade – can lead to disastrous consequences.

Despite institutional shortcomings and other economic difficulties, the importance of economic integration among developing countries as a policy option for fostering development and overcoming the constraints of small domestic markets has been already recognized (Bhagwati and Panagariya 1996; Schiff 1996). Integration into the regional economy may also be seen as a 'stepping stone' to future integration into the world economy. However, a key question for developing and transition economies is the model or approach to follow in pursuing this gradualist approach to greater integration in the global economy.

One approach for developing and transition economies is to pursue integration among neighbours at equivalent levels of economic development, progressively undertaking liberalization as their economies develop, and deepening the integration process beyond the frontiers. This may be seen as an intermediate step for developing countries towards the full implementation of WTO commitments, balancing the lower benefits of fuller integration against the adjustment process.

Another option for developing and transition countries is to form or join 'mixed' RTAs in partnership with a developed country, perhaps with different transition periods, but with both sides assuming basically similar obligations for the longer term. In principle, mixed groupings with major trading partners should provide improved stability of access to product and factor markets for developing countries than those available from sub-regional groupings with neighbouring developing countries as well as enhanced investment and growth (Whalley 1996). The developed country partner may also provide finance or other support for its partners in such grouping, as in the case of the Europe and Euro-Med Agreements. Apart from enhanced market opportunities and investment flows, an agreement with a developed country is more likely to lead to deeper integration and a stronger legal and institutional framework for trade, benefiting national producers and traders as well as partners (World Bank 2000). Under NAFTA, Mexico was able to expand both its trade and investment to the US and Canada in the first year of its membership in NAFTA, but locking in domestic reforms was a prime motivation from the Mexican side (GATT 1993). This was also the case in the EU – Turkey customs union (Hartler and Laird 1999). Cyprus and Malta also experienced the rapid expansion of their exports to the EU in the first years of their RTAs.

Apart from the legal analysis of actual RTA provisions and their consistency with WTO rules, another way of analysing the effects of South–South

RTAs is to gauge their actual impact on trade flows among members, and between members and third countries: in the final analysis the multilateral system is not an end in itself but the means of promoting economic progress among members (see preamble to the WTO Agreement). A vast literature discusses the trade and welfare effects in great theoretical and empirical detail. One typical yardstick applied to any RTA is whether the overall effect is *trade creation* or *trade diversion à la* Viner. From this perspective, a large number of South–South RTAs do not seem to be more trade diverting than North–North RTAs. Cernat (2001), for instance, used a gravity model to estimate the impact of South–South RTAs on both intra- and extra-trade flows. Unlike widespread opinions and standard theoretical predictions, the empirical evidence suggests that several South–South RTAs (such as COMESA, ASEAN, CARICOM) are not trade diverting but trade creating, both with regard to intra- and extra-RTA trade.[23] For instance, the empirical estimates suggest that two COMESA members traded in 1998 2.6 times as much as otherwise similar countries. At the same time, trade between COMESA members and third countries increased by 25 per cent as a result of COMESA formation (Cernat 2001). What these findings suggest is that even though some of these South–South RTAs faced implementation problems and delays in liberalizing intra-regional trade, the formation of these RTAs succeeded in removing some 'invisible' trade barriers between members despite the absence of major tariff preferences. This may well be from the reduction of non-negligible transport costs, border formalities, technical or health standards, and other measures that are captured by what is referred to as 'trade facilitation' measures may also impose significant costs.[24] All these 'invisible' cost-increasing elements may all be reduced through the formation of a South–South RTA. Eliminating such trade barriers implies no welfare loss since there are no tariff revenues forgone (Baldwin 1994).

Another form of 'trade facilitation' effect of RTA formation in the case of African RTA is put forward by Glenday (1997). He argues that, in theory, RTAs can strengthen intra-regional cooperation among African countries to promote intra-regional trade and to allow more efficient border controls through sharing of import documents; common control systems should make circumvention less attractive (the Lafer concept of lower taxation increasing revenues). Such intergovernmental cooperation can also render corruption and red tape more difficult.

While these arguments assist in understanding the estimated results, two additional questions are raised by this explanation based on 'invisible' trade costs. First, can RTAs be held accountable for this outcome? Second, do RTAs eliminate these trade barriers in a discriminatory manner, so as to explain the wedge between gross trade creation and diversion estimates? The answer to the first question can be found in the objectives of most South–South RTAs. Most of these trading arrangements involved regional cooperation in a number of areas with direct relevance for trade patterns: upgrading transport and communication, infra-structure, harmonization and simplification of custom procedures, trade facilitation measures for transit goods, and so on. Such objectives and concrete initiatives have been carried out, more or less successfully, by many of the South–South RTAs.

With regard to the second question, whether the elimination of such 'invisible' trade costs induces discrimination between RTA members and third countries is less straightforward. One can easily distinguish the complex set of regional initiatives aimed at fostering trade in discriminatory and non-discriminatory policies. Given the weak implementation record of most South–South RTAs, immediately after their formation tariff reductions on intra-RTA trade are far from universal. Yet, the RTA formation appears to reduce some of the non-tariff barriers on both intra-RTA trade and even on third country exports to the region.[25] One can imagine a number of other costs that differently affect intra- and extra-regional exports, whose removal will introduce an implicit differentiation in total trade costs. The overall effect will be a slightly larger reduction of trade barriers on intra-RTA trade (both tariff and non-tariff reduction) compared to non-members (only some 'invisible' non-tariff barriers reduced).

In sum, even though they are in most of the cases less ambitious in their achievements than North–North RTAs, several South–South RTAs could serve as 'building blocks' for more open trade in line with WTO objectives. Backsliding from multilateral commitments may still occur, but the existence of various forms of RTAs (North–North, South–South, or mixed) can act as regional stabilizers for trade policy formation and as a policy transfer mechanism from regional to multilateral level. Such a regional policy transfer mechanism could advance the WTO agenda to include issues of specific concerns for developing countries. Such regional groupings could also contribute to a more effective participation of developing countries in multilateral negotiations through coordinated negotiation positions (as in the case of MERCOSUR on market access negotiations or SADC on services, for instance).

## Conclusion

The objective of this chapter was to address the question as to whether RTAs contribute to the smooth functioning of the multilateral trading system, in particular to what extent RTAs can prevent protectionist 'backslidings'. More specifically, the chapter looked in detail at a number of trade policy instruments (anti-dumping, subsidies and countervailing duties, safeguard measures, technical barriers to trade and standards, customs procedures and rules of origin) and trade-related policies (trade-related investment measures, competition policy, movement of labour), as well as dispute settlement mechanisms in the context of regional agreements. We also looked at different models or configurations of RTAs to try to see how these complement the multilateral trade disciplines and avoid the use of 'soft' WTO rules for protectionist purposes.

As the evidence presented in this chapter suggests, although several well-advanced North–North RTAs have put in place various policy-stabilizing mechanisms, they are unevenly applied across RTAs and across trade issues. Therefore, the risk of certain trade policy backsliding among RTA members through the use of anti-dumping, countervailing measures and TBTs still remains, in particular where the degree of integration is low. Another notable aspect of advanced North–North RTAs is that when the policy stabilizing mechanisms do not apply to non-RTA members the potential for trade-diverting backsliding actions against third partners is greatly increased.

Even more uneven stabilizing capabilities can be found in South–South RTAs. With few exceptions, South–South RTAs are 'shallower' than their North–North counterparts. Furthermore, for those 'deep' integration schemes among developing countries, the implementation status does not yet match the ambitious objectives set out in the preambles of their RTA agreements. This is largely related to economic and institutional factors that have led to a rather cautious approach to market opening, as well as to the potential adjustment costs. Another aspect that may explain the implementation gap of certain South–South RTAs is that, unlike North–North or North–South RTAs, regional schemes among developing countries do not need to fulfil the rather strict requirements imposed by GATT Article XXIV on the design and implementation of regional agreements. South–South RTAs are usually notified at the WTO under the Enabling Clause, which offers a great deal of flexibility and leaves greater room for implementation gaps to RTA members in terms of the scope and pace of regional integration. However, unlike North–North RTAs, less advanced South–South RTAs do not make large scale use of 'postmodern' trade distorting tools such as anti-dumping and countervailing duties, neither against RTA partners nor against third countries. Nor do they make much use of dispute settlement mechanisms to challenge other countries when they become a target of such measures.

Compared to South–South RTAs, mixed RTAs have been praised by some authors for their potential 'lock-in effects', issues of implementation capacity, asymmetry, reciprocity and traditional concerns about particularly sensitive product sectors render the negotiation of mixed agreements difficult. Furthermore, not all mixed RTAs aim at producing a more level playing field than the multilateral trading system. 'Postmodern', hidden protectionist trade instruments, such as anti-dumping practices, are still present even in advanced North–South or East–West RTAs, such as the Europe Agreements. They become even more acute when potential developing partner countries have a large production capacity for sensitive products, such as staple foods, fruit and vegetable products, clothing or textiles, that are protected by tariff peaks and escalation or other non-tariff barriers in the developed market.[26]

Overall, despite the lack of detailed evidence on the operation of many agreements, our examination of the provisions, a reading of the WTO examination process and a range of studies suggest that RTAs have anticipated the WTO in a number of areas and continue to do so, particularly where there are lacunae in the regulatory framework or weak WTO provisions. There are cases, however, when multilateral trade negotiations have influenced the shape of regional integration schemes, in particular with regard to those issues that were brought recently on the WTO agenda.[27] In recent years, changes have been most marked in behind-the-borders measures – deeper integration – and current discussions on investment and competition policy indicate that this process is ongoing.

While there is an argument that such deeper integration is beneficial in increasing the security and predictability of trading conditions, some developing countries in particular feel that pressures to extend WTO rules is a strain on their capacity to absorb and implement as well as limiting the flexibility they have to pursue their own developmental policies. The relatively slow pace of South–South agreements reflects this concern as well as concerns about adjustment costs in much the same way as many view trade liberalization in general.

The scope and pace of South–South RTAs has led some commentators to favour the 'North–South' model, in which a developed country or group of countries act as the anchor for an agreement. Apart from the improved market opportunities (and perhaps other assistance), this model may also be seen as a way of forcing the pace and expanding the scope of RTAs and ensuring that what is agreed is fully implemented. Some developing countries have themselves seen this as an advantage in locking in their own reforms.

While RTAs represent a weakening of the MFN principle in practice there is not much evidence from quantitative studies of serious trade diversion and there are cases where third countries seem to benefit. One major exception to this seems to be the application of rules of origin where there are currently no WTO disciplines, a gap that needs to be remedied.

To some extent the new initiatives within RTAs reflect impatience with the slow process of multilateral negotiations. Countries and businesses that want to move faster are able to do so within a regional framework. The pressure is then on multilateral trading systems to move in these areas to reduce the scope for discrimination, and by and large the system has proved its capacity for further, gradual extension. However, the way in which further elaboration of the multilateral trading system takes place is critical. It could provide developing and transition countries with important advantages for defending their interests vis-à-vis partners with stronger bargaining power. It could also increase the pressures on them to take on new commitments for which they are not yet ready.

# Notes

1 For surveys of the literature on this debate, see Bhagwati and Panagariya (1996), Laird (1999) and Winters (1996). Other authors examining the issue of whether the formation of regional arrangements leads to higher or lower protection with respect to the outside countries, see Bagwell and Staiger (1993), Levy (1997), and Panagariya and Findlay (1996).

2 For an interpretation of GATT Article XXIV's requirement that RTAs eliminate 'other restrictive regulations of commerce', see, for instance, Mavroidis (1997). The crucial question is which regulations of commerce are actually deemed to be restrictive. However, it is argued that anti-dumping and countervailing duties are simply defensive instruments aimed to reduce the negative impact of other restrictive policies such as dumping or subsidies.

3 See, for instance, Article 30 of Protocol VIII, amending the treaty establishing the Caribbean Community.

4 Although the general application of safeguards is strengthened under NAFTA, the agreement provides for special safeguards on textile and clothing products.

5 Argentina – Safeguards Measures on Imports of Footwear, Report of the Panel (WT/DS121/R, 25/06/1999) and Report of the Appellate Body (WT/DS121/AB/R, 14/12/1999).

6 For a comprehensive discussion of this case, as well as a broader analysis of safeguards and regionalism, see Mathis (2002).

7 The exclusion of Canada is in accordance with Article 1102 of the CUSFTA, which provides for mutual exclusion from global safeguard actions under GATT Article XIX unless imports from the other Party were 'substantial' and 'contributing importantly' to the serious injury or threat thereof caused by increased imports. The CUSFTA standards in respect of emergency safeguard actions were essentially carried over into NAFTA.

8 In accordance with the Agreement on Safeguards, developing countries accounting for less than 3 per cent of the US imports were also excluded from the safeguard measures.

9 This exclusion of RTA partners from the applicability of US steel safeguard measures has been challenged by Japan as a violation of the MFN principle (Article I of the GATT 1994 and Article 2.2 of the WTO Agreement on Safeguards) and by Korea as a violation of Articles 2, 3, 4 and 5 of the WTO Agreement on Safeguards.

10 EFTA states protested against their inclusion under the EU measures and argued that the exclusion of products originating in the EEA states is permitted under the WTO, provided that such imports are also excluded from the injury determination, and provided such non-application is necessary under the free trade agreement. However, under the EEA, safeguards are permitted under Articles 112–114, which state that safeguard measures can be applied between EEA partners 'if serious economic, societal or environmental difficulties of a sectoral or regional nature liable to persist are arising'.

11 This provision was used by Chile on imports of four agricultural products (wheat, wheat flour, sugar and edible oils). However, Canada was excluded from the extended measure for sugar and vegetable oils, but not for wheat or wheat flour. The safeguard for wheat and wheat flour was subsequently lifted following protests by Canada (based on information available at www.dfait-maeci.gc.ca).

12 Based on information available at www.ecowas.int.

13 Based on information available at http://www.sadcreview.com/.

14  Controls through a single, shared physical infrastructure in which the neighbouring countries' customs services operate side-by-side.

15  For instance, as the Canada – Auto Pact case has demonstrated, despite the prohibition of mandatory performance requirement, NAFTA did not exclude the maintenance of voluntary performance requirement associated with duty waivers. Such non-mandatory provisions were *de facto* deemed to be discriminatory.

16  This initiative resembles the EU initiative on the European Company Statute, which allows companies to be incorporated at European, as opposed to national, level.

17  The EEA agreement was preceded by a series of bilateral FTAs between the EC and EFTA countries. These 'first generation' agreements contained provisions on competition policy that relied on the Joint Committee for dealing on anticompetitive practices. These provisions were subsequently reproduced in many other agreements between the EU and other countries in the region, and between various candidate countries.

18  See the WTO website for a list of notified RTAs that contain provisions on services liberalization.

19  For a recent survey of regional liberalization of trade in services in East Asia and the Western Hemisphere, see, for instance, Nikomborirak and Stephenson (2001).

20  See WTO (2002a) for a comprehensive survey of ROOs contained in more than 90 RTAs notified to the WTO.

21  For a description of the COMESA rules of origin, see in particular Rule 2 of Annex 1, Protocol on the Rules of Origin for Products to be Traded between the Member States of the Common Market for Eastern and Southern Africa.

22  This is particularly true in the case of NAFTA where several companies offer online specialized services on how to fulfil the NAFTA ROOs requirements to interested producers.

23  Similar results are reported by Frankel (1997) and Winters and Wang (1994) on ASEAN, Boisso and Ferrantino (1997) for CARICOM. Primo Braga, Safadi, and Yeats (1994) found positive trade creating effects for CACM.

24  Hoekman and Konan (1999) found compelling evidence of such 'invisible' costs. Thus, according to them, only redundant testing and idiosyncratic standards alone imposed extra-costs from 5 per cent to 90 per cent of the value of traded goods.

25  Hartler and Laird (1999) note that third parties benefited from the EU – Turkey customs union as the trade regime became more open and enforcement more predictable.

26  See UNCTAD (2000) and Cernat, Laird and Turrini (2002) for detailed analyses of the tariff peaks and escalation faced by developing countries. On the persistence of tariff peaks on sensitive products in RTAs, see WTO (2002b).

27  For instance, the negotiation of the certain Europe Agreements between the EU and Central and East-European countries were influenced by the progress achieved in the Uruguay Round of multilateral negotiations.

# References

Aninat, A. (1996) *RTAs: Options for Latin America and the Caribbean*, UNCTAD Study prepared for the Seminar on Regional Economic Arrangements and their Relationship with the Multilateral Trading System, Geneva 1996.

Bagwell, K. and R. Staiger (1993) 'Multilateral Tariff Cooperation During the Formation of Regional Free Trade Areas', NBER Working Paper no. 4364.

Baldwin, R. (1994) *Towards An Integrated Europe*, London: CEPR.

Bhagwati, J. and A. Panagariya (1996) 'Preferential Trading Areas and Multilateralism – Strangers, Friends, or Foes?', in J. Bhagwati and A. Panagariya (eds),*The Economics of Preferential Trade Agreements*, Washington, DC: AEI Press.

Boisso, D. and M. Ferrantino (1997) 'Economic and Cultural Distance in International Trade: An Empirical Puzzle', *Journal of Economic Integration*, vol. 12, no. 4, December.

Bora, B. and C. Findlay (1996) *Regional Integration and the Asia-Pacific*, Melbourne: Oxford University Press.

Bora, B., P.J. Lloyd and M. Pangestu (2000) 'Industrial Policy and the WTO', *The World Economy*, vol. 23, no. 4, pp. 543–59.

Cernat, L. (2001) 'Assessing Regional Trade Arrangements: Are South–South RTAs More Trade Diverting?', *Global Economy Quarterly*, vol. II, no. 3 (July–September).

Cernat, L. (2002) 'Assessing Regional Trade Agreements: Same Issues, Many Metrics', UNCTAD Policy Issues in International Trade and Commodities, New York and Geneva: United Nations.

Cernat, L., S. Laird and A. Turrini (2002) 'How Important are Market Access Issues for Developing Countries in the Doha Agenda?', Centre for Research in Economic Development and International Trade, Nottingham: University of Nottingham CREDIT Research Paper no. 02/13.

Crawford, J.A. and S. Laird (2000) 'Regional Trade Agreements and the WTO', Nottingham: University of Nottingham, CREDIT Research Paper No. 00/3.

De Rosa, D.A. (1998) 'Regional Integration Arrangements: Static Economic Theory, Quantitative Findings, and Policy Guidelines', World Bank Policy Research Report *Regionalism and Development*, Background Paper no. 2007.

Devlin, R., A. Estevadeordal, L.J. Garay (1999) 'The FTAA: Some Longer Term Issues', Buenos Aires, INTAL ITD Occasional Paper no. 5.

Drabek, Z. and S. Laird (1998) 'The New Liberalism: Trade Policy Developments in Emerging Markets', *Journal of World Trade*, vol. 32, no. 5, October.

Foroutan F. (1993) 'Regional Integration in SSA: Past Experience and Future Prospects', in J. de Melo and A. Panagariya (eds), *New Dimensions of Regional Integration*, Oxford: Oxford University Press.

Frankel, J.A. (1997) *Regional Trading Blocs in the World Economic System*, Washington, DCL: Institute for International Economics.

GATT (1993) *Trade Policy Review – Mexico 1993*, Geneva.

Glenday, G. (1997), 'Customs and Trade Facilitation: Challenges and Opportunities in Sub-Saharan Africa', Harvard Institute for International Development, Development Discussion Paper no. 616.

Hartler, C. and S. Laird (1999) 'The EU Model and Turkey – A Case for Thanksgiving?' *Journal of World Trade*, vol. 33, no. 3, June.

Hoekman, B. (1998) 'Free Trade and Deep Integration: Antidumping and Antitrust in Regional Agreements', Washington, DC: World Bank, World Bank Policy Research Working paper.

Hoekman, B. and D.E. Konan (1999) 'Deep Integration, Nondiscrimination, and Euro-Mediterranean Free Trade', Washington, DC: World Bank, World Bank Policy Research Working Paper no. 2130.

Kerr, W. (1997) 'Removing Health, Sanitary and Technical Non-Tariff Barriers in NAFTA', *Journal of World Trade*, vol. 31, no. 5, 57–73.

Kerr, W. (2001) 'Greener Multilateral Pastures for Canada and Mexico: Dispute Settlement in the North American Free Trade Agreements', *Journal of World Trade*, vol. 35, no. 6, 1169–80.

Laird, S. (1999) 'Regional Trade Agreements – Dangerous Liaisons?', *The World Economy*, vol. 22, no. 9, December.

Lawrence, R.Z. (1996) *Regionalism, Multilateralism and Deeper Integration*, Washington, DC: Brookings Institution.

Levy, P. (1997) 'A political-economic analysis of free trade agreements', *American Economic Review*, no. 87, 506–19.

Mathis, J. (2002) *Regional Trade Agreements in the GATT/WTO: Article XXIV and the Internal Trade Requirement*, The Hague: TMC Asser Press.

Mavroidis, P. (1997) 'Comments', in P. Demaret *et al.* (eds), *Regionalism and Multilateralism after the Uruguay Round*, Brussels: European Interuniversity Press, pp. 389–96.

Mikesell, R.F. (1963) 'The theory of common markets as applied to regional arrangements among developing countries', in *International trade theory in a developing world: Proceedings of a conference held by the International Economic Association*, R. Harrod and D. Hague (eds), New York: St Martin's Press.

Musonda, J. (2000) 'Enhanced Technical Assistance and Co-operation: Priorities for the Effective Implementation of Competition Policy in Africa – the Case of COMESA', Paper presented at the WTO Regional Workshop on Competition Policy, Economic Development and the Multilateral Trading System, Cape Town, South Africa, February.

Niels, G. and A. ten Kate (1997) 'Trusting antitrust to dump antidumping: abolishing antidumping in free trade agreements without replacing it with competition law', *Journal of World Trade*, vol. 31, no. 6, 29–43

Nikomborirak, D. and S. Stephenson (2001) 'Liberalization of Trade in Services: East Asia and the Western Hemisphere', Paper presented at the PECC Trade Policy Forum, Bangkok, June 12–13.

Nogues, J. and R. Quintanilla (1993) 'Latin America's Integration and the Multilateral Trading System', in J. de Melo and A. Panagariya (eds), *New Dimensions in Regional Integration*, Cambridge (UK): Cambridge University Press.

Panagariya, A. and R. Findlay (1996) 'A Political-Economy Analysis of Free Trade Areas and Customs Union', in R. Feenstra, D. Irvin, and G. Grossman (eds), *The Political Economy of Trade Reform*, Essays in Honor of Jagdish Bhagwati, Cambridge, Mass: MIT Press.

Primo Braga, C., R. Safadi and A. Yeats (1994) 'Implications of NAFTA for East Asian Imports', Washington, DC: World Bank, World Bank Policy Research Working Paper no. 1351.

Rose, A. (2002) 'The Effect of Common Currencies on International Trade: Where do we stand?', mimeo, Haas School of Business, University of California, Berkeley, CA, (Draft, August 7).

Schiff, M. (1996) 'Small is beautiful: Preferential trade arrangements and the impact of country size, market size, efficiency, and trade policy', Washington, DC, World Bank, Policy Research Working paper.

UNCTAD (2000) 'The Post-Uruguay Round Tariff Environment for Developing Country Exports: Tariff Peaks and Tariff Escalation', TD/B/COM.1/14/Rev.1, Geneva.

United Nations Economic and Social Commission for Asia and the Pacific (UNESCAP) (1998) 'Implications of the APEC Process for interregional trade and investment flows', *Studies in Trade and Investment*, no. 33, Bangkok: United Nations.

Whalley, J. (1996) 'Why do countries seek regional trade agreements?', NBER Working Paper no. 5552.

Winters, L.A. (1996) 'Regionalism versus Multilateralism', Washington, DC, World Bank, Policy Research Working Paper.

Winters, L.A. and Z.K. Wang (1994) *Eastern Europe's International Trade*, Manchester: Manchester University Press.

World Bank (2000) *Trading Blocs*, Oxford: Oxford University Press.

WTO (2002a) *Rules of Origin Regimes in Regional Trade Agreements*, Geneva: WTO, 17 September, WT/REG/W/45.

WTO (2002b) *Coverage Liberalization Process and Transition Provisions in Regional Trade Agreements*, Geneva: WTO, 5 April, WT/REG/W/46.

WTO (2002c) *ANZCERTA Free Trade Agreement between Australia and New Zealand: Biennial Report of the Operation of the Agreement*, Geneva: WTO, 13 May, WT/REG111/R/B/2.

Yeats, A.J. (1998) 'What Can Be Expected from African Regional Trade Arrangements? Some Empirical Evidence', World Bank Working Paper no. 2004.

# Part II
# Trade and Macroeconomic Policy in Transition Countries

# 3

# Trade Policy and Macroeconomics in European Transition Economies

*David Begg*

## Introduction

Economic transition is about getting control of macroeconomic policy and making progress on the supply side through structural adjustment. Prospective membership of, first, the EU and, then, the EMU itself will affect both aspects of transition. This clarity about eventual goals is helpful. However, there is a danger that transition economies are urged into the premature adoption of policies that will be appropriate only later. Assessing how policy regimes should evolve requires an interesting characterization of what is meant by transition itself.

In this chapter I focus on three themes. The first is stabilization of public finances, and hence credible policies for steady disinflation. The second is investment in costly structural adjustment to build the institutional capacity necessary to remove distortions and expand aggregate supply as transition proceeds. The third is a concern about external solvency and development of the traded goods sector, and hence concern about the allocation of expenditures between domestic and external sectors, from which we may deduce the appropriate path of competitiveness during transition. Within this framework, I then consider the second-best case for trade distortions (for example, tariffs) that enhance competitiveness and generate valuable tax revenue, but come at the cost of supply-side distortions. I show that in some circumstances, these trade measures usefully supplement macroeconomic policy design; in other circumstances, their cost outweighs their benefit.

Many of us have written at length about desirable and undesirable aspects of macroeconomic policy design in transition economies. For a recent discussion of its relation to EMU, see Begg, Halpern, and Wyplosz (1999). For example, there is increasing recognition of the danger of attempting to stabilize the exchange rate within a narrow band, both because this policy is vulnerable to speculative attack as capital mobility increases. This danger is often exacerbated by misguided attempts to sterilize capital inflows, thereby increasing inflation and enhancing the vulnerability of the economy – notably, of its banking system – to a change in foreign sentiment, and hence a subsequent capital outflow.

This lesson, advocated early in Begg (1996), has been well learned, not least because of the lessons drawn from the subsequent Asian crises. Thus, countries such as the Czech Republic and Poland have abandoned pegged exchange rates in favour of domestic nominal anchors in which inflation targets play a prominent role. Their exit from pegged regimes, and its contrast with the maintenance of a narrow crawling band in Hungary, is discussed further in Begg and Wyplosz (1999).

Floating exchange rates place a safety valve on the extent of capital inflows, thereby helping insulate transition economies from subsequent crises. Even so, domestic anchors carry their own dangers. First, since Dornbusch (1976) economists have appreciated that disinflation through tight domestic monetary policy may cause exchange rate overshooting. In OECD economies with a long tradition of markets and of competition, and in which banks are well regulated and have deep pockets, coping with exchange rate misalignment may be costly but not disastrous. In transition economies, any lengthy period of uncompetitiveness will inflict larger costs, and may also provoke a domestic backlash in favour of greater protection.

Reform is costly today but improves future opportunities. I capture cumulative progress to date by the extent to which an effective tax base has been established and enforced. Although it is only one of many aspects of supply side improvement, development of tax capacity is a natural link between structural adjustment of the supply side and the public finance constraint on fiscal policy, monetary policy and inflation.

Begg (1996) analyses monetary and fiscal regimes as the delegation of operational powers in response to market distortions or commitment failures. The delegation of monetary policy is the subject of a large literature beginning with Rogoff (1985), and has led to the creation of many independent central banks, often in conjunction with the inflation targets recommended in Svensson (1997).

In contrast to monetary delegation, the literature on the delegation of fiscal powers is tiny, though in practice the finance ministry is usually delegated to an 'Iron Chancellor', the fiscal analogue of a conservative central banker. However, finance ministers are fired more easily than central bank governors. Even so, attempts at fiscal delegation are what lie behind both the budget conditions imposed by external agencies such as the IMF, and the Stability Pact in the EU.

Given interactions between monetary and fiscal policy, it is inappropriate to design monetary institutions that neglect failures in fiscal policy. Yet, most of the literature on monetary institutions ignores this reality.[1] Transition economies in the 1990s faced many problems of commitment, including the difficulty in sticking to plans to set adequate tax rates (including dispensing with subsidies at the planned rate), and the difficulty of implementing promises to undertake structural adjustment. Begg (1996) discusses implications for the chosen speed of structural adjustment and the forms of institutional design that mitigate these problems.

Here, I assume that such problems have already been solved by prior institutional design, and focus on the role of trade policy. One object of future research should be to investigate how the current results are altered if other policy failures – in monetary policy, fiscal policy or commitment to reform – have not yet been addressed.

I explore a small open economy whose structure evolves slowly over time. Aggregate demand depends on competitiveness, the real interest rate and fiscal policy. Government spending is financed partly by the inflation tax and partly by distortionary taxes that reduce output. There are two taxes. Although tariffs are distortionary in the sense of reducing total output, they boost demand for traded goods. Other taxes are distortionary at any point in time only when set in excess of the economy's current tax capacity. At the start of transition, countries have a low tax capacity. They respond by choosing high levels of distortionary taxes (and correspondingly low output), high inflation and low government spending. As transition proceeds, tax capacity improves, allowing for lower levels of distortionary taxes and decreasing the need for inflation tax revenue to finance government spending. Thus, inflation falls and output rises. This does not reflect a Phillips curve but rather the fact that high tax distortions and high inflation are both optimal responses to a low inherited tax capacity. Structural adjustment improves all options for the future. In contrast, undue tightening of monetary policy deprives the government of the (optimal) amount of inflation, creating further reliance on distortionary taxes that depress output unnecessarily.

I then endogenise the supply side. Structural reform has costs and benefits. Benefits are the ability to raise more tax revenue for any given level of output distortion. However, I assume more rapid reform is increasingly costly, which may have economic or political interpretations. The optimal speed of reform is derived. Within this framework, I examine the interaction of monetary policy, fiscal policy, trade policy and the chosen speed of structural adjustment.

## A small open transition economy

Consider a small transition economy. Distortionary taxes reduce equilibrium output, but the degree of distortion depends on the size of the tax base and the degree of tax compliance. Suppose $t^+$ describes the 'tax capacity' of an economy. Distortions arise when actual taxation $t$ exceeds $t^+$. Structural adjustment increases tax capacity $t^+$, allowing the government to finance larger spending levels in a nondistortionary way.

To focus on competitiveness and trade policy, I neglect shocks and policy surprises. I also set many parameters, but not all, equal to unity so that there is less algebra to carry around. Generalizations are straightfoward. I assume output supply obeys

$$y = -\tau - \bar{\omega} \quad \tau = t - t^+, \quad \tau, \omega \text{ nonnegative} \tag{1}$$

Long run output is normalized to zero. In the short run, output is depressed by excess domestic taxation $\tau$ and by tariffs $\omega$.

In the long run, I will allow the government to invest in enhancing $t^+$; for a given tax rate $t$ this reduces $\tau$ and raises aggregate supply $y$. This also makes clear that the analysis is much more general than the 'tax capacity example' with which I work. If the reader prefers, $t^+$ can be viewed as *any* structural determinant of potential output in which the government can invest; for example, infrastructure or regulatory expertise. Equation (1) is then reinterpreted as stating that tax and tariff rates $t$ and $\omega$ reduce output, but the relevant structural capital stock $t^+$ increases output. For linguistic convenience, I continue to refer to $t^+$ as tax capacity in what follows. Whether a failure of tax compliance is the most pressing supply side issue or not varies from country to country. It has obviously been a greater problem in Russia and the Ukraine than in the Czech Republic or Hungary. Even in these latter countries, there are many other candidates for important supply-side variables for which $t^+$ can be viewed as a proxy. Structural adjustment is then a costly investment in increasing $t^+$

The government supplies goods and services $G$, financed by the two taxes and by the inflation tax

$$G = t + \omega + \pi \tag{2}$$

Following Debelle and Fischer (1994), the inflation-tax Laffer curve is linear in the relevant range, and for simplicity I take its slope to be unity.

Aggregate demand depends on competitiveness, which is determined by $c$ (the logarithm of the real exchange rate) and by tariffs $\omega$; on the forward-looking real interest rate (which by interest parity equals $c_{+1} - c$), and on government spending $g$

$$y = [c + \omega] - b[c_{+1} - c] + jG \quad 0 < j < 1 \tag{3}$$

Although the financing of $G$ is assumed to reduce aggregate demand, the government budget constraint holds in each period. I therefore assume that the *net* demand effect of an increase in $G$ is positive, though less than unity.

I shall shortly describe the government's loss function which includes a time-invariant target level $G^*$ for government spending. One aspect of my characterization of transition is therefore that the government is ambitious, in the sense that it would like to achieve the steady state level of government spending even in early transition when tax capacity is low and output supply is consequently highly distorted. It is convenient to define $g = G - G^*$, whence (3) may be rewritten

$$y = [s + \omega] - b[s_{+1} - s] + jg \quad g = G - G^*, \ s = c + jG^* \tag{4}$$

It is helpful to rewrite (2) as

$$g = \tau + \omega - h + \pi, \qquad h = G^* - t^+ > 0 \tag{5}$$

Crucially, $G^*$ exceeds the initial capacity $t^+$ for nondistortionary taxation. The government uses both excess taxation and the inflation tax to move initial levels of government spending closer to its target level. Over time, as tax capacity rises, tax distortions and inflation can fall. Indeed, once $t^+$ eventually increases to $G^*$, all other variables fall to zero.

The per period loss function of the government is

$$L = \pi^2 + y^2 + g^2 + (s + \omega)^2 + \mu[h - h_{-1}]^2 \quad \mu > 0 \tag{6}$$

Despite inheriting a low tax capacity, the government cares about deviations of inflation, output, government spending, and competitiveness from their steady state levels – all of which will turn out to be zero. Since equation (1) implies that short run equilibrium output is $-[\tau + \omega] < 0$, a transition economy will experience an output recession as the government strives for high levels of tax revenue to pursue its spending ambitions.

The final term in (6) captures the cost of structural adjustment, which may be economic or political. Here, I take the simplest case in which there is simply an increasing marginal cost of reform. This of course will generate sluggish adjustment to smooth the costs of reform. Early in reform, poor credibility may justify rapid reform to take advantage of a window of opportunity. However, most European transition economies are now more securely established on continuing reform, for which steady adjustment is a more plausible description. For a further discussion of credibility issues and their effect on reform, see Begg (1996). Note, too, that it would be possible to allow reform costs to depend on other variables within the model.

## The solution

I now wish to solve the intertemporal optimization problem in choosing interest rates, the two tax rates, and the speed of structural adjustment.[2] Because of interest parity, the current real exchange rate depends on its own expected future values, raising the issue of time inconsistency in policy design. It might be possible to precommit all policies in a fashion that rules out the temptation to use future surprises. Here, I assume such perfect precommitment is impossible. Market particpants therefore anticipate that future governments will reoptimize and build this into initial expectations; formally, the solution must impose time consistency by using dynamic programming.

In this model, prices are flexible and inspection of the equations of the previous section shows there is only one predetermined variable, $h_{-1}$. The general form of the solution is therefore easy to characterize. Along the rational expectations saddlepath, $h$ will evolve according to

$$h = \lambda h_{-1} \qquad 0 < \lambda < 1 \tag{7}$$

and all other endogenous variables will be linear in $h$ (and hence in $h_{-1}$ too). The time consistent solution chooses the contemporaneous value of the policy variables, which we can think of as $(\tau, \omega, s, h)$, knowing that their future values will depend on current choices through dynamics such as (7).

Letting $s_{+1} = \lambda s$, substituting this and (5) into (4) and rearranging

$$j\pi = -[1+j]\tau - [2+j]\omega + jh - [1+\beta(1-\lambda)]s \tag{8a}$$

which expresses the endogenous variable $\pi$ in terms of the four policy parameters. For convenience, I restate how the other endogenous variables depends on these parameters

$$y = -\tau - \omega \tag{8b}$$

$$g = \tau + \omega - h + \pi \tag{8c}$$

I assume that the government minimizes the present value of losses $[L + \phi L_{+1}]$ subject to equations (8a)–(8c) and using the expression for $L$ in equation (6).

Differentiating respectively with respect to $s, \omega, \tau$, and $h$

$$0 = [\pi + g][-1 - \beta(1-\lambda)]/j \qquad +[s+\omega] \tag{9a}$$

$$0 = [\pi + g][-2 - j]/j \quad -y + g \qquad +[s+\omega] \tag{9b}$$

$$0 = [\pi + g][-1 - j]/j \quad -y + g \tag{9c}$$

$$0 = [\pi + g] \qquad -g \qquad +\mu(h - h_{-1}) - \phi\mu(h_{+1} - h) \tag{9d}$$

Solving equations (9a)–(9c)

$$0 = [s+\omega] = [\pi + g] = g - y \tag{10}$$

whence

$$\pi = -y = -g = [\tau + \omega] = h/3 = \lambda h_{-1}/3 \tag{11}$$

As conjectured, these variables are linear in $h$, and hence linear in the state variable $h_{-1}$. Initially, inherited $h_{-1}$ is positive, reflecting the tensions between fiscal aspirations and structural adjustment to date. The government's optimal tradeoff entails some inflation, some output recession, and some distortionary taxes to finance some government spending, though less than the government's ideal level in the long run. Two issues remain. The chosen speed of adjustment, and the implications tariff policy.

## Tariff policy

Consider the tariff policy for a given rate of structural adjustment $\lambda$. From (8) and (11)

$$jh/3 = j\pi = \omega\beta(1 - \lambda) - (1 + j)(\tau + \omega) + jh = \omega\beta(1 - \lambda) + [h/3][-1 - j + 3j]$$

whence

$$3\omega = h(1 - j)/[\beta(1 - \lambda)] \tag{12}$$

For all $j < 1$ in equation (3), equation (12) confirms that positive tariffs $\omega$ are part of the optimal policy along the transition. Equation (12) suggests that the tariff rate $\omega$ is linear in $h$, and falls smoothly towards zero as $h$ converges on zero as a result of gradual structural adjustment.

However, it is necessary to check what is happening to other taxes $\tau$. Since $[h/3] = [\tau + \omega]$ equation (12) then implies

$$[3\beta(1 - \lambda)]\tau = [\beta(1 - \lambda) - (1 - j)]h \tag{13}$$

which yields positive excess taxes $\tau$ if, and only if

$$\beta(1 - \lambda)] > (1 - j) \tag{14}$$

which requires that the rate of structural adjustment is reasonably rapid (small $\lambda$) and that fiscal policy has a large effect on output ($j$ close to 1) even when aggregate demand effects of fiscal financing are fully recognized. The latter is most likely when the private sector faces considerable capital market imperfections in borrowing, as may be plausible during the early stages of transition.

In the event that (14) is satisfied, the complete implications for trade policy have now been derived, and can be explained intuitively. Suppose the government made no use of tariff policy so that $\omega = 0$ throughout transition. It would no longer be possible to ensure that competitiveness remained at the (zero) level considered appropriate in the government's loss function. Using equations (8), (9a) and (9c), it can then be shown

$$\tau = \pi = g = -y = h/3 \tag{15}$$

and thus, using (8a)

$$s[1 + \beta(1 - \lambda)] = (1 - j)y < 0 \tag{16}$$

Hence, in the absence of tariffs, the economy is uncompetitive along the optimal transition path ($s < 0$), although this problem unwinds as the steady state approaches, since $s$ is also linear in $h$.

Relative to this scenario, the additional ability to use tariffs, despite distorting total output, gives the government an additional policy weapon with which it can redress uncompetitiveness. Indeed, as I have set up the base case, tariffs are better than excess domestic taxes since they yield the same marginal output distortion and same marginal financing of government spending but additionally can help alleviate uncompetitiveness.

Beginning from zero tariff levels, the transition economy therefore prefers to raise tariffs a little, and reduce other taxes a little, achieving a better external position for the same total output and government spending. Tariffs allow the government to manipulate the allocation of private demand between foreign goods and domestic goods. If other (excess) taxes are still positive once $(s + \omega)$ has been driven to zero, that is the end of the story.

However, it is possible that domestic excess taxes $\tau$ fall to zero before $\omega$ has risen to the level that achieves $(s + \omega) = 0$. The mathematics of equations (9)–(15) then supposes that negative levels of domestic excess taxation would actually boost domestic output supply. However, this is probably unrealistic. Rather, the output equation (1) also entails a nonnegativity constraint on excess taxes.[3] Thus when condition (14) fails, complementary slackness applies in condition (9c). Optimal policy then sets $\tau = 0$ throughout transition – actual taxes $t$ grow exactly in line with tax capacity $t^+$ as transition proceeds – and equations (8)–(9) are then solved subject to this restriction (which gives a different answer).

Consider the marginal incentive to raise tariffs further once excess taxes have now been driven to zero. In particular, consider further revenue neutral switches reducing taxes $t$ and increasing tariffs $\omega$. The marginal benefit is the effect on $(s + \omega)$ though $s$ is endogenous and partly offsets changes in tariffs. The marginal cost was previously zero (since excess taxes and tariffs had equal distortionary effects on output supply). However, once the corner solution is reached, further reductions in marginal benefit $t$ yield no benefit whereas further increases in $\omega$ continue to depress output supply. There is therefore a discrete jump (from zero to one unit of output) in the marginal cost of revenue-neutral switches in taxation towards tariffs. If $(s + \omega)$ is already close to zero, the effect of further tariff increases may therefore be outweighed by the marginal cost.

Whether we are in the full interior solution or at this corner solution, the optimal tariff policy sets positive tariff rates, linear in $h$, that shrinks to zero as the steady state approaches. Having understood how trade policy is set, I turn finally to the dynamics of transition itself.

## Structural adjustment

When there is an interior solution for $\tau$, equation (10) implies that $\pi + g = 0$. Moreover, along the time consistent path, assuming $h = \lambda h_{-1}$ at all *future* dates must then make it optimal to choose $h = \lambda h_{-1}$ for the current $h$ as well. Hence, equations (9d) and (10) then imply $0 = h/3 + \mu(h - h_{-1}) - \phi\mu(h_{+1} - h) = h_{-1}[\lambda/3 + \mu(\lambda - 1) - \mu\phi\lambda(\lambda - 1)]$
whence

$$0 = f(\lambda) = -\phi\lambda^2 + \lambda[(1 + \phi) + (1/3\mu)] - 1 \tag{17}$$

Since $f(0) < 0 < f(1)$, this confirms that there is a unique convergent root $\lambda^*$ which is a positive fraction, as Figure 3.1 illustrates; the other root of the quadratic is $\lambda^{**} > 1$, which we can ignore since it is explosive. Note too that since an increase in adjustment costs $\mu$ shifts the function $f(\lambda)$ downwards for all positive values of $\lambda$, this confirms that a more costly adjustment then slows down the optimal rate of reform by increasing the persistence parameter $\lambda$.

How would the rate of transition be affected if, instead of this interior optimum, excess taxes had been set at the zero corner solution? Since the only purpose of tariffs is to increase competitiveness $(s + \omega)$ from its negative value in the absence to tariffs towards the optimal value of zero in the full interior solution, being forced to stop short must leave $(s + \omega) < 0$.

Notice that equations (9a), (9b) and (9d) are invariant across these regimes. In the corner solution regime with $\tau = 0$, (9a) therefore implies that $(\pi + g) < 0$ since $(s + \omega)$ must be negative in this regime. Hence, in general, the values of all endogenous variables are affected. The optimal second best response is to spread the burden around – a little total tax revenue, a little government

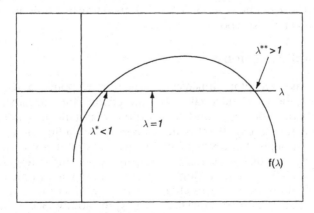

*Figure 3.1* The optimal rate of persistence $\lambda^*$

spending, and much more to inflation to prevent government spending from falling too much.

It is the bit of extra inflation that gives the clue to the answer in the corner solution case. Suppose

$$\pi = h(B + 1/3) \qquad B \text{ nonnegative} \tag{18}$$

where $B = 0$ in the interior solution (11). Hence (9d) may now be rewritten

$$
\begin{aligned}
0 &= [\pi + g] \qquad - g \qquad + \mu(h - h_{-1}) - \phi\mu(h_{+1} - h) \\
&= h_{-1}[f(\lambda) + \lambda B]
\end{aligned}
\tag{19}
$$

where f $(\lambda)$ is given in (17). Examining Figure 1, any addition $\lambda B$ shifts the function upwards during the interval in which $\lambda$ is positive has the consequence of reducing the optimal second-best degree of persistence by reducing $\lambda$. This has a simple intuitive explanation. For given $h_{-1}$, the marginal cost of adjustment is independent of the other economic variables but the marginal benefit is higher the more distorted the economy, since convergence to the long run equilibrium is more rapid. Hence, there is an incentive to reform more quickly when first-best tariff policy cannot be pursued because excess domestic taxes have already been driven to zero.

Notice that this also precludes a multiple equilibrium arising through dependence of the regime selection criterion (14) on the speed of reform $\lambda$. Had we concluded that persistence was higher in the corner solution regime, it might have been possible to argue that conjecturing high persistence would make $\beta(1 - \lambda)$ small, cause inequality (14) to fail, and validate belief in the corner solution; whereas conjecturing low persistence would make $\beta(1 - \lambda)$ large, validating belief in the interior solution. However, we have seen that the effects go in the opposite direction – the corner solution regime has the lower persistence. There will therefore be a unique conjecture, validating the analysis above.

## Lessons and extensions

My brief was to explore linkages between trade policy and macroeconomic policy design. I have taken the protection embodied in tariff rates as my index of trade policy and analysed the simultaneous choice of trade policy, monetary policy, fiscal policy and structural adjustment. These are connected through the government budget constraint, by a concern for price stability, and by concern about competitiveness and external solvency. Meaningful analysis of transition economies must contain an interesting role for transition itself. In this chapter, the government undertakes costly investment to gradually improve tax capacity, the ability to levy and enforce broad-based taxes with low or zero distortionary effect.

Within the framework, the inherited amount of structural adjustment is the key state variable describing the economy's current and future possibilities. With a poor inheritance, an economy faces horrible tradeoffs which make the optimal policy entail high inflation, low government spending, highly distortionary taxes and low competitiveness. Cross country correlations – for example, the famous negative correlation of output and inflation even in the longer run (see, e.g., Bruno and Easterly, 1995) – should properly be interpreted not as a causal relationship but as joint symptoms of a common underlying problem, the poor structural inheritance. It may, therefore, be completely inappropriate to recommend ultra tight monetary policy in the hope of reducing inflation and boosting output. Within the current model, forcing inflation below its optimal path merely forces some reliance on additional taxes that are more distortionary and some acceptance of even lower levels of government spending than should be warranted.

Within my stylized framework, tariffs are, by construction, a better tax than domestic taxes, in that they do everything domestic taxes can do but also enhance (suboptimally low) competitiveness. In the absence of tariffs, competitiveness is initially too low but gradually climbs towards zero. Interest parity means that the real exchange rate is gradually depreciating from an initially high level. Since high real interest rates are part of the optimal strategy of disinflation and transition, this is unavoidably an aspect of the first-best path. However, trade policy can make a difference, by cushioning the economy from the initial real appreciation.

Provided domestic excess taxes are not driven to zero, competitiveness – determined jointly by the real exchange rate and the level of protection – can be maintained at its ideal level by a judicious choice of trade policy. However, if the marginal distortion caused by domestic taxes falls to zero once taxes fall to tax capacity, the device of revenue neutral switches from domestic taxes to tariff revenue can no longer improve competitiveness at zero marginal cost. This regime may therefore be forced to stop short of driving competitiveness all the way to its ideal level. If so, facing higher distortions than in the interior solution, reform will be more rapid since the payoff will be larger.

Why, then, might one feel uncomfortable with the policy recommendation to proceed with an aggressive policy of tariffs? Five reasons suggest themselves. One is the question of retaliation. If failure to participate rapidly in liberal international trade regimes induces foreign retaliation, exports may be reduced, upsetting the *ceteris paribus* assumptions behind the previous analysis.

Second, it is well known that, having installed a regime of tariffs, it is then politically difficult to remove them. In terms of modelling strategy, this suggests that it may be as important to keep track of the costs of tariff reduction as it is to keep track of the costs of expanding domestic tax capacity. Conceptually, this poses no serious problems for the analysis, but it introduces a second state variable – the inherited level of tariffs. With two predetermined variables, the saddlepoint will now need two convergent roots. No problem for the computer to solve but no longer conducive to easy analytical solution as in Figure 3.1. Nevertheless, investigating this case is a high priority for future research.

Third, one might wish to extend the analysis by making the costs of domestic reform also depend on aspects of the state of the economy. For example, if reform takes real resources that displace other items of government spending, it would then be more costly to reform when a poor inheritance already constrains options for government spending. In a linear-quadratic model, optimal policies will still be linear functions of the two inherited state variables, which will converge steadily on the steady state. In models with greater nonlinearity, however, it might then be possible to justify initial periods of inaction – waiting for sufficient improvement in circumstances before tackling a really difficult issue – or initial bouts of intense activity.

Fourth, in this chapter have I emphasized the normal commitment problems in monetary policy (though my time consistent solution precludes any attempt to surprise the currency markets). Since one theme of my analysis is that all decisions are optimally simultaneous, any failures in other aspects of policy will affect the choice of trade policy and of reform speed. Begg (1996) has addressed many of these issues within a closed economy, examining not merely how monetary policy independence may counteract the temptation to inflation but also how commitment problems may arise, and may be overcome, in relation to both fiscal policy (the use of surprise lump sum taxes that avoid output distortions) and reform (promising, but not delivering on, costly reform in order to try to induce beneficial private behaviour *ex ante*).

Fifth, to make the points of the chapter dramatically, I assumed that the marginal supply-side distortion of tariffs was no higher than the marginal supply-side distortion of general taxation once its tax capacity threshold had been reached. However, this need not be the case. Beginning from zero tariff levels, the marginal benefit of tariffs is essentially the benefit of a marginal improvement in competitiveness, whereas the marginal cost is zero if the extra tariff revenue is used to reduce other excess taxes so that the net marginal effect on output distortions is zero. If, however, the marginal output distortion of tariffs is sufficiently larger than that of other taxes, even at small tariffs levels, then it may be optimal to choose a corner solution in which tariff levels are zero. More generally, a larger marginal output distortion caused by tariffs reduces the optimal tariff rate and reduces the likelihood that large tariffs are part of optimal policy design even early in transition.

Finally, the framework I have developed allows one to address issues such as future regime switches to trade union (EU) and monetary union (EMU). One could think of the former as a jump to zero tariffs at some future date and the latter the jump to an exogenously nominal interest rate coupled with a fixed nominal exchange rate: changes in domestic inflation – for example, caused by fiscal policy – would thus affect both real interest rates and the real exchange rate. One could thus compute both policy and welfare after such a regime had been entered, but also the effects on policy in the earlier regime in which exchange rate expectations correctly anticipate subsequent behaviour. Since future regime changes would be capitalized into contemporary competitiveness, this would clearly allow a role for trade policy to offset any undesirable effects. However, if entry to the EU or EMU depended on prior attainment of satisfactory values of both policy and endogenous market variables, this would constrain policy choices in further interesting ways.

There are, therefore, many ways in which the analysis of this chapter could be taken forward. My claim is not to have solved all these issues linking the design of trade policy with the design of monetary and fiscal policy and the chosen pace of structural adjustment. However, I hope that the framework that I have developed not merely sheds insights on simple versions of these issues, but also helps point the way for further research.

## Notes

1  Exceptions include Beetsma and Uhlig (1997), Beetsma and Jensen (1999), and Debelle and Fischer (1994).
2  Sibert (1998) is an early integration of macroeconomic policy and incentives for structural adjustment.
3  Notice that when the interpretation of $t^+$ is generalized from tax capacity to *any* determinant of potential output, there is no longer any need to worry about negative values of $t - t^+$ (and hence negative $\tau$).

## References

Beetsma R. and H. Jensen (1999) 'Structural convergence under reversible and irreversible monetary policy', CEPR Discussion Paper 2116.

Beetsma R. and H. Uhlig (1997) 'An analysis of the Stability Pact', CEPR Discussion Paper 1669.

Begg, D. (1996) 'Monetary policy in Central and Eastern Europe: lessons after half a decade of transition', IMF Working Paper WP/96/109a.

Begg, D. and C. Wyplosz (1999) 'Untied hands are fundamentally better', paper presented at the 5th Dubrovnik Conference, June 1999, forthcoming in M. Blejer and M. Skreb (eds), *Ten Years of Transition*, Cambridge (UK): Cambridge University Press.

Begg, D., L. Halpern and C. Wyplosz (1999) 'Exchange rate adjustment in transition Economies', *Economic Policy Initiative 5*, Centre for Economic Policy Research.

Bruno, M. and W. Easterly (1995) 'Inflation crises and long-run growth', NBER Working Paper 5209.

Debelle, G. and S. Fischer (1994) 'How conservative should a central bank be?', in J. Fuhrer (ed.), *Goals, Guidelines and Constraints facing Monetary Policymakers*, Federal Reserve Bank of Boston.

Dornbusch, R. (1976) 'Exchange rate dynamics', *Journal of Political Economy*, vol. 84, no. 6 (December).

Dornbusch. R. and S. Fischer (1993) 'Moderate inflation', *World Bank Economic Review*, vol. 7, no. 1 (January).

Rogoff, K. (1985) 'The optimal degree of commitment to an intermediate monetary target'; *Quarterly Journal of Economics*, vol. 100, no. 4 (November) pp. 1169–89.

Sibert, A. (1998) 'Monetary convergence and economic reform', *Economic Journal*, vol. 109, no. 452 (January), pp. 78–92.

Svensson, L. (1997) 'Optimal inflation targets, conservative central banks, and linear inflation contracts', *American Economic Review*, vol. 87, no. 1 (March) pp. 98–114.

# 4
# Czech Trade, Exchange Rate and Monetary Policies in the 1990s

*Michaela Erbenova and Tomas Holub**

## Introduction

This chapter concentrates on the interaction between Czech trade and sta-bilization policies in the 1990s. The authors argue that the initial transition package was a coherent one in which stabilization policies supported both internal and external liberalization of the economy. Later on, however, macroeconomic imbalances started to emerge, aggravated by institutional weaknesses and massive capital inflows in 1993–96. The central bank's ster-ilization policies proved to have little effect. Moreover, the fiscal policy was rather pro-cyclical. This led to a current account crisis in early 1997, followed by an enforced switch to floating and a painful stabilization period. The authors analyse the causes of this crisis and the stabilization policy mix using a simple econometric model of the current account. The crisis challenged the liberal approach to trade policy, leading to some minor backsliding in trade policy, such as the introduction of an import deposit scheme. However, the Czech government has relied primarily on pro-export measures rather than anti-import measures in its more activist trade policy.

## Non-technical summary

In this chapter, we concentrate on the interaction between trade policy and stabilization policies throughout the 1990s. We focus mainly on those periods when macroeconomic policies had an indirect negative impact on the Czech trade, on the competitiveness of Czech industry, and in particular on the current account crisis of 1995–97.

* This research was undertaken with support from the European Union Phare ACE Programme project number P97-8129-R, directed by Dr Zdenek Drabek. We thank David Begg, Zdenek Drabek and Zdenek Tuma for helpful comments and sugges-tions. Any remaining errors and omissions, however, are our own. Views expressed in this paper do not represent official views by the Czech National bank.

We argue that the initial transition package was a consistent one, in which the restrictive stabilization policies strongly supported the main reform goals (fast macroeconomic stabilization, far reaching liberalization and westward reorientation of foreign trade). From 1994 to mid-1997, however, the policy of fixed nominal exchange rate, accompanied by sterilized interventions and coupled with a pro-cyclical fiscal policy, proved to be unsustainable. Moreover, these policies created a favourable environment for short-term capital inflows, which further softened the already weak budget constraints in the economy.

The structural and institutional weaknesses of the Czech economy played an adverse role in 1994–99. They limited the range of sustainable growth rates far below the originally expected levels, thus contributing to macroeconomic overheating. The worsening current account deficit was a primary reflection of this development. We use a simple econometric analysis of the current account in this chapter to simulate developments that would have been sustainable. We conclude that a sustainable scenario, consistent with the EU-convergence requirements, would have been characterized by an annual GDP growth of 2.5–3.5 per cent, a CPI-based real exchange rate appreciation of 1.2–2.2 per cent a year and a stable PPI-based real exchange rate. The actual performance during 1994–96 was much different than these values.

In 1997–99, a correction took place characterized by a decline in GDP, moderation in the appreciation of real exchange rate and in the reduction of the current account deficit. This correction was the result of restrictive monetary policy measures in mid-1996, and of further macroeconomic restrictions in the spring of 1997. Our analysis suggests that the correction was not excessive given the economic fundamentals, but it lead to a significant deviation from the optimal path towards the EU-convergence. The optimal adjustment path should have been more growth- and trade-friendly and it should have relied more on a stimulation of exports through an exchange rate depreciation rather than on a reduction of imports generated by a sharp restriction of domestic demand.

Such an optimal adjustment path was probably not fully achievable in practice, but the actual performance would have been better with a different policy mix. The monetary policy tightening of mid-1996 slowed down the growth of aggregate demand (with a lag), but also supported the exchange rate appreciation in late 1996, having an ambiguous short-run impact on the current account. Instead, a fiscal tightening would have been preferable, but 1996 was an election year and exactly the opposite happened. The problem thus can be viewed as a coordination failure between fiscal and monetary policies. This is not to say, however, that monetary policy was optimal. The CNB should have probably tried to soften the tight monetary conditions by larger interest rate cuts, and possibly also by more active foreign exchange interventions in 1998. At that time, there was no domestic reason to let the monetary conditions tighten to their historical maximum, and thus to deepen and lengthen the economic recession. On the other hand, it was feared in 1998 that the emerging market crisis might lead to a quick reversal in foreign capital flows. This, in turn, serves as an explanation for the slow pace of interest rate cuts in 1998.

At present, a key question for the central bank is how to treat the exchange rate in the new inflation-targeting regime. In 1999–2000, the CNB appears to have been concerned about the appreciation of exchange rate, as documented by foreign exchange interventions and the creation of special privatization accounts. Nevertheless, the CZK has been appreciating quite rapidly due to fast FDI inflows. Another important macroeconomic risk at present is the rising fiscal deficit, which is of structural origin and will require a substantial adjustment in the future, with potentially severe short-term consequences for the economy.

From the trade policy's point of view, the rising trade and current account deficit in 1995–97 brought a challenge to the liberal doctrine in foreign trade. As a result, the focus of trade policy started to shift towards more activism by the second half of the 1990s. In 1995, the Czech Export Bank was established that provides export financing to Czech companies. While the growing current account deficit was not the primary stimulus for the creation of this institution, it surely increased political support and speeded up the process. Also, the Export Guarantee and Insurance Corporation has intensified its operation since 1996.

Three anti-import measures were enacted during the current account crisis. First, an anti-dumping law was adopted in June 1997. Second, a shift in trade policy focus was reflected in the government recommendation to give a preferential treatment to domestic suppliers in public procurement. And finally, an import deposit scheme was introduced in the spring of 1997 obliging importers of selected goods to deposit 20 per cent of the imported goods value on a non-interest-bearing account for 180 days. Import deposits represented a clear policy slippage with an adverse impact on the Czech Republic's reputation. Moreover, the measure proved to be ineffective in macroeconomic terms, and was abolished in August 1997.

In spite of these anti-import measures, it can be argued that the Czech Republic has relied primarily on export promotion as its major trade policy tool since 1997. The new government committed itself, in 1998, to create a coherent pro-export policy based on development of the Czech Trade Agency, Export Guarantee and Insurance Corporation and Czech Export Bank. A considerable expansion of both financial and non-financial export promotion activities has occurred since 1999. Financial measures include, most importantly, financing and underwriting of state-supported exports, targeted aid for development, and the system of the so-called 'soft loans', that is, long-term government loans on favourable terms. The government has also emphasized the need for greater role of trade diplomacy. Nevertheless, it is likely that, despite the sharp increase in financial support, these measures influence the structure of exports rather than the total volume of exports and the trade balance.

The chapter is organized as follows. The first section provides a brief historical overview of the macroeconomic developments and policy measures during the 1990s. The next section presents an econometric analysis of the Czech current account. Its results are then used to assess the stabilization policies in the concluding section.

## A historical overview

In this section, we provide a historical overview of the economic performance in the 1990s. We start with a short general description of the whole period, which is followed by a more detailed overview of monetary policy and trade policy.

### A short general overview

The Czech Republic (and, formerly, Czechoslovakia) had both favourable and unfavourable starting conditions compared to other post-communist countries at the beginning of economic transition. The advantages included a history of reasonably low inflation, both open and suppressed,[1] relatively small monetary overhang, low foreign debt[2] and traditionally prudent fiscal policy with net government debt in 1989 representing less than 1 per cent of GDP. Among the disadvantages, there was an almost complete state control over the economy (including prices, foreign trade and so on), domination of markets by large scale, heavy industry enterprises and the monopoly of trade and a heavy orientation of foreign trade on post-communist countries (see, for example, Dyba and Švejnar 1995). Less than 0.5 per cent of non-agricultural output was produced in the private sector in 1990. Almost 60 per cent of Czech exports went to the East in 1988, while less than 35 per cent was directed to advanced market economies (see Table 4.4). Czechoslovakia was more dependent on socialist trade than other Central and East European countries, except for Bulgaria. The collapse of CMEA[3] thus presented a greater danger for the country than for any of its post-socialist neighbours.

The major policy goals of the early period of transition included a fast macroeconomic stabilization, far reaching liberalization (including foreign trade liberalization) and westward reorientation of foreign trade. Some preliminary steps had been taken already during 1990, including a sharp devaluation of the exchange rate and an abolition of the negative rate of the turnover tax, which was equivalent to a dramatic scaling down of subsidies. The main reform package was then introduced at the beginning of 1991. The currency was declared 'internally convertible', liberalizing most current account transactions, as well as some capital account transactions (the inflow of foreign direct and portfolio investments, repatriation of profits and so on). Foreign trade restrictions (quotas, tariffs and so on) were substantially reduced during 1991. The majority of all prices were freed in January 1991, leading to a one-off jump in the price level in early 1991. The monetary policy was designed as strongly restrictive in the first two years of transition in order to prevent this initial price jump from generating subsequent inflationary spirals.

The restrictive monetary policy measures, together with the breakdown of the CMEA and inherited structural problems of the economy, caused a deep recession during which the Czech Republic lost almost 15 per cent of the starting GDP (see Table 4.1).

The economy started to show the first signs of an economic revival during 1992, but the subsequent period was affected by the split up of the Czechoslovak Federation on January 1, 1993, which contributed to a stagnation of GDP in 1993. Moreover, this political shock coincided with the introduction of a new tax system[4] that pushed the inflation rate above 20 per cent in 1993 (see Table 4.1).[5]

The years 1994–96 can be characterized as a period of fast GDP growth, excessive wage growth (see Table 4.1), an appreciating real exchange rate due to an inflation differential and massive inflows of foreign debt finance (see Table 4.2) associated with the capital account liberalization. All these factors contributed to a mounting current account imbalance (see Table 4.2).

The CNB tried to fight these unfavourable trends by widening the fluctuation band of the exchange rate in February 1996 and by introducing a set of restrictive monetary policy measures in the second half of 1996. These measures, however, proved to be insufficient. In May 1997, the Czech Republic experienced a currency turmoil that forced the CNB to float the exchange rate and intensify its restrictions. In addition, fiscal restrictions were introduced in the spring of 1997. This led to a period of a painful stabilization, marked by a drop in GDP and sharply rising unemployment in 1997–99, a rapid disinflation (Table 4.1) and a reduction in the current account deficit (Table 4.2).

The years 2000–01 again witnessed a revival in economic growth. The recovery was also accompanied by a turnaround in inflation and trade deficit. These were caused primarily by the external oil price shock and a food

*Table 4.1* Basic economic indicators (per cent)

|  | 1990 | 1991 | 1992 | 1993 | 1994 | 1995 | 1996 | 1997 | 1998 | 1999 | 2000 |
|---|---|---|---|---|---|---|---|---|---|---|---|
| GDP growth | −1.2 | −11.6 | −0.5 | 0.1 | 4.8 | 5.9 | 4.3 | −0.8 | −1.2 | −0.4 | 2.9 |
| Inflation | 9.7 | 56.6 | 11.1 | 20.8 | 10.0 | 9.1 | 8.8 | 8.5 | 10.7 | 2.1 | 3.9 |
| Wage growth | 3.7 | 15.4 | 22.5 | 25.3 | 18.5 | 18.5 | 18.4 | 10.5 | 9.4 | 8.2 | 6.6 |
| Unemployment | 0.7 | 4.1 | 2.6 | 3.5 | 3.2 | 2.9 | 3.5 | 5.2 | 7.5 | 9.4 | 8.8 |

*Source*: Czech Statistical Office, Czech National Bank.

*Table 4.2*   Czech balance of payments (CZK bn)

|  | 1993 | 1994 | 1995 | 1996 | 1997 | 1998 | 1999 | 2000 |
|---|---|---|---|---|---|---|---|---|
| Current account | 13.3 | −22.6 | −36.3 | −116.5 | −101.9 | −43.1 | −54.2 | −87.7 |
| of which: Trade balance | −15.3 | −39.8 | −97.6 | −159.5 | −144.0 | −82.4 | −65.8 | −120.8 |
| Financial account | 88.2 | 97.0 | 218.3 | 113.6 | 34.3 | 94.3 | 106.6 | 130.0 |
| of which: FDI | 16.4 | 21.6 | 67.0 | 34.6 | 40.5 | 115.9 | 215.7 | 172.8 |
| Portfolio inv. | 46.7 | 24.6 | 36.1 | 19.7 | 34.4 | 34.5 | −48.3 | −68.2 |
| Other capital | 25.1 | 50.9 | 115.2 | 59.2 | −40.6 | −56.1 | −60.9 | 26.9 |
| Errors and omissions | 3.0 | −6.1 | 15.8 | −19.6 | 11.2 | 11.3 | 4.9 | −10.5 |
| Change in FX reserves* | −88.3 | −68.3 | −197.9 | 22.5 | 56.0 | −62.6 | −57.1 | −31.6 |

*Note*:   * '−' means an increase in reserves.

*Source*:   Czech National Bank.

price hike, even though the recovery in domestic demand has started to play an increasingly important role as well since late 2000. These developments, however, were reversed again in the second half of 2001 due to low commodity prices during the global slowdown.

### Monetary policy overview

In order to liberalize the foreign trade regime at the beginning of transition, a more realistic exchange rate was needed. There was a strong uncertainty about the level of the equilibrium exchange rate. The devaluation had to be large enough to ensure that domestic price increases would absorb the monetary overhang without endangering both real devaluation and export competitiveness. At the same time, the devaluation should not jeopardize macroeconomic stability by fuelling higher inflation. The authorities preferred a radical devaluation, possibly meaning an initial overshooting, to avoid the potential danger of an overvalued currency and future devaluation spirals. The Czechoslovak crown was devalued in four steps (two in January, one in October and one in December 1990), overall by 113.4 per cent,[6] and then pegged to a basket of the currencies of five major trading partners.[7]

The sharp exchange rate devaluation together with price liberalization in January 1991 pushed the inflation rate to 56.6 per cent in 1991 (up from 10 per cent in 1990).[8] The monetary policy was designed to be highly restrictive in the first two years of transition in order to prevent the initial price jump from producing subsequent inflationary rounds. In 1990, the nominal volume of domestic credit and money supply, in fact, stagnated. The discount rate was raised several times during 1990 (from 4 to 8.5 per cent). In early 1991, the restrictions further intensified. The discount rate was raised to 10 per cent (on 28 December, 1990). The growth of money supply reached 27.3 per cent in 1991 (see Table 4.3), leading to a substantial drop in the real quantity of money. In fact, the actual size of the monetary restriction was

*Table 4.3* Basic monetary indicators (in per cent)

| | | 1990 | 1991 | 1992 | 1993 | 1994 | 1995 | 1996 | 1997 | 1998 | 1999 | 2000 |
|---|---|---|---|---|---|---|---|---|---|---|---|---|
| M2 growth | e-o-y | 0.1 | 27.3 | 20.3 | 19.8 | 19.9 | 19.8 | 9.2 | 10.1 | 5.2 | 8.1 | 6.5 |
| **Interest rates** | | | | | | | | | | | | |
| –discount | e-o-y | 8.5 | 9.5 | 9.5 | 8.0 | 8.5 | 9.5 | 10.5 | 13.0 | 7.5 | 5.0 | 5.0 |
| –Pribor 3M | a-o-y | n.a. | n.a. | 12.7 | 13.1 | 9.1 | 11.0 | 12.0 | 16.0 | 14.3 | 6.8 | 5.4 |
| –deposits | a-o-y | n.a. | 8.1 | 6.7 | 7.0 | 7.1 | 7.0 | 6.8 | 7.7 | 8.1 | 4.5 | 3.4 |
| –credits | a-o-y | n.a. | 14.5 | 13.5 | 14.1 | 13.1 | 12.8 | 12.5 | 13.2 | 12.9 | 8.7 | 7.2 |
| –new credits | a-o-y | n.a. | n.a. | n.a. | 14.9 | 13.0 | 13.2 | 13.4 | 16.2 | 14.7 | 8.6 | 6.9 |
| **Exchange rate (against former parity)** | | | | | | | | | | | | |
| –parity=100 | a-o-y | 63 | 101 | 100 | 100 | 100 | 100 | 99 | 105 | 106 | 111 | 113 |
| –y/y nominal | a-o-y | n.a. | –61 | 0.3 | 0.7 | 0.1 | 0.0 | 0.9 | –6.5 | –0.8 | –4.5 | –2.3 |
| –y/y real CPI | a-o-y | n.a. | –6.5 | 6.2 | 14.6 | 6.7 | 6.3 | 7.1 | –0.3 | 7.9 | –3.4 | –0.4 |
| –y/y real PPI | a-o-y | n.a. | 3.7 | 8.0 | 11.7 | 4.5 | 5.3 | 4.8 | –2.6 | 4.4 | –3.3 | –0.3 |

*Source:* Czech Statistical Office, Czech National Bank, author's computations.

*Note:* Data prior to 1993 refer to Czechoslovakia; a negative sign means a depreciation of the exchange rate.

greater than intended due to a cautious lending policy of commercial banks and ceilings on credit expansion and refinancing credits introduced by the State Bank (see Klacek and Hrnčíř 1994).

The monetary policy contributed substantially to a stabilization of price level in mid-1991.[9] Thanks to this, the initial boost to price competitiveness due to the exchange rate devaluations was not fully eroded by inflation.[10] Competitiveness was also supported by a fall in real wages in 1991, largely attributable to the government's system of wage controls. As shown in Table 4.1, the nominal wages increased by only 15.4 per cent, compared to the inflation rate of 56.6 per cent. Combined with a reduction in domestic demand, the above developments led to a reduction in imports and to an unexpectedly fast territorial reorientation of exports (see Table 4.4). As a result, the current account shifted into a surplus in 1991.

It was often asked whether the initial exchange rate devaluation and monetary restrictions were not excessive. In *ex-post* terms, this opinion may be correct, as the unexpected current account surplus of 1991 indicates. But given the extent of *ex-ante* uncertainties, we can only conclude that the initial package of reform measures was consistent and that the stabilization policies strongly supported the main reform goals, including the objectives of trade policy (i.e. external stabilization, trade liberalization, reorientation of exports).

During 1992, the central bank started to loosen its monetary policy. The discount rate was cut overall by 1.5 percentage points from January to August 1992, credit limits were removed, and the money supply growth substantially exceeded the rate of inflation in 1992 (see Table 4.3). At the turn of 1993, however, the policy easing was interrupted by the split of the Czechoslovak Federation (combined with a jump in the price level of about 8 percentage points due to the new tax system) which had a strong adverse impact in the monetary sphere. The investors feared a devaluation of the currency, and this contributed to a decline of foreign exchange reserves of the Czech National Bank (CNB) to just USD 0.5 bn. in January 1993. The CNB fought these negative developments hard. The discount rate was increased again from 8.0 per cent to 9.5 per cent at the end of 1992, the minimum reserve ratio was raised in February 1993, and open market operations were used to withdraw liquidity from the banking sector. As a result, the growth of money supply slowed down to about 15 per cent year-on-year in the first half of 1993 (see Figure 4.1), and the market interest rates increased (see Table 4.3 and Figure 4.3).

After the separation of the Czech and Slovak currencies in March 1993, the situation quickly stabilized. The foreign exchange reserves began to grow again (to USD 3.8 bn in December 1993). The inflation rate, which jumped up sharply in January 1993, returned to its gradually declining trend (see Figure 4.2). As a result, the CNB loosened its monetary policy again in mid-1993.

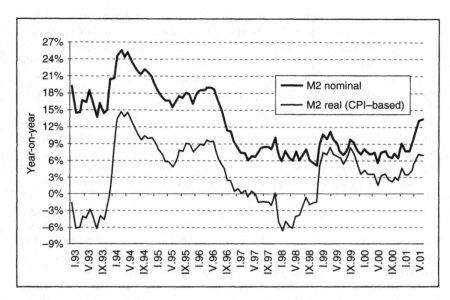

*Figure 4.1* Growth of money supply

*Source*: Czech National Bank.

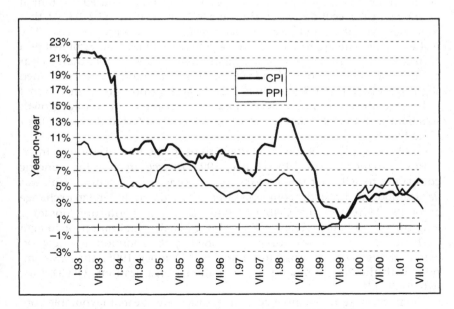

*Figure 4.2* Consumer price and producer price inflation since 1993

*Source*: Czech Statistical Office.

To reduce the need of additional foreign exchange reserves after the currency separation and to facilitate the initial mutual balance of payments adjustments, a temporary payment agreement between the Czech and Slovak Republics was put in place between March 1993 and October 1995. The transactions between the two countries were settled through a special clearing account operated by the two central banks, the clearing exchange rates of the two currencies being pegged to the ECU. There was a possibility of devaluation or revaluation against the clearing ECU of up to 5 per cent. In early March 1993, the Czech Republic revalued by 2 per cent and the Slovak Republic devaluated by 5 per cent (but revalued later on). As a result, there was in fact a dual exchange rate for mutual transactions between the Czech and Slovak Republics as compared to their transactions with other countries, which was unfavourable for the Czech Republic.

The exchange rate level of the CZK against the basket of reference currencies remained fixed at its previous level, and thus continued to serve as a 'nominal anchor'; that is, a tool for importing low inflation and credibility from abroad. In real terms, however, the exchange rate started to appreciate quickly from 1992 (see Table 4.3) due to the inflation differential. Combined with an excessive wage growth (see Table 4.1), that was not justified by adequate increases in labour productivity, this began to erode the price competitiveness boosted at the beginning of transition by the exchange rate devaluations. The unbalanced tendencies, which later on caused substantial problems in the economy, had thus already emerged in 1992–93. At that time, however, the stabilization policies seemed to be quite coherent. The real exchange rate appreciation and wage growth could be interpreted as a correction of the initial undershooting in these variables rather than a clear origin of problems to come.[11] Indeed, the current account balance remained slightly positive in 1993 (see Table 4.2), indicating no significant external imbalance. However, the relatively rapid inflation and nominal wage growth was built into the expectations and thus adversely affected the subsequent period.[12]

In 1994–95, the CNB took important steps in the liberalization of international financial flows.[13] In February 1994, the requirement for domestic companies to sell their foreign exchange earnings to the central bank was *de facto* abolished, and the companies were allowed to hold foreign exchange accounts. Moreover, the CNB was quite liberal in permitting domestic enterprises to obtain foreign credit. This gradual liberalization of foreign exchange transactions was formally finished by the adoption of a new foreign exchange law that came into force on 1 October, 1995. This law complied with the basic standards of international organizations (OECD, IMF, Article VIII).[14]

The liberalization of capital account transactions created favourable conditions for a massive inflow of foreign debt finance to the Czech economy in 1993–96 (see Table 4.2). The fast inflow of short-term capital was composed of portfolio investment and foreign borrowing by Czech banks and companies. The Czech banking sector played a particularly important role in channelling foreign debt finance into the economy, its net foreign borrowing peaking at 6 per cent of GDP at the end of 1995. In the fixed exchange rate regime, this caused a strong upward pressure on money supply. Even though the central bank tried to sterilize its foreign exchange interventions, the sterilization was becoming less effective over time (see Frait 1996; Šmídková 1996). There was thus little the CNB could do to stabilize the economy, as the autonomy of its monetary policy was seriously constrained by the exchange rate regime. In this respect, the growth of money supply was rather fast (see Figure 4.1), persistently exceeding both the nominal GDP growth and the CNB's targets (see Hrnčíř and Smídková 1998; Jonáš 1996). Moreover, as the subsequent developments showed, the allocative efficiency of the Czech banking sector was not particularly impressive. Many loans were provided to large companies in traditional industries that later on got into insolvency problems, and this contributed to further deterioration in the economic performance. In other words, the capital inflows not only increased the overall liquidity in the economy, but may have also been a factor that helped further soften the already weak 'budget constraints'. As a result, the inflation rate was declining only gradually in that period (see Figure 4.2), and it was thus still well above the inflation rate in advanced market economies, causing a

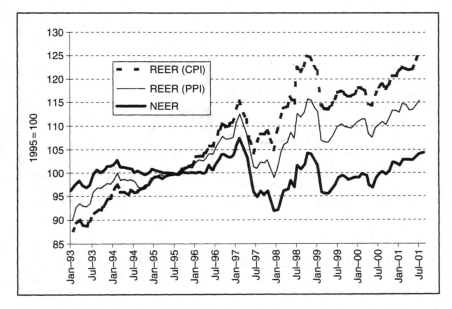

*Figure 4.3* Nominal and real effective exchange rate
*Source:* Czech National Bank.

continuous real exchange rate appreciation (Figure 4.3).[15] The growth of real wages also remained high, thus further weakening the price competitiveness of Czech companies (see Table 4.1).[16]

In order to regain control over monetary aggregates, domestic demand and current account, the CNB widened the band of exchange rate fluctuation to ±7.5 per cent in February 1996. This contributed to an outflow of short-term capital, which, combined with the high current account deficit, turned the overall balance of payments into a deficit for the first time since 1993 (see Table 4.2). Moreover, the CNB raised all of its major interest rates (by 1.0–1.5 percentage points) and the minimum reserve requirement (from 8.5 per cent to 11.5 per cent) in order to slow down the domestic credit creation in mid-1996. The market interest rates went up, both in nominal and real terms (see Figure 4.4), and the growth of money supply substantially slowed down during the second half of 1996 (see Figure 4.1). The higher interest rates, however, contributed to a nominal strengthening of the exchange rate (see, for example, Čihák 1997), which accelerated the appreciation in real exchange rate (see Figure 4.3.). The monetary restrictions thus probably had a negative impact on the current account in the short run, as the influence of exchange rates on foreign trade tends to be faster than the effect of monetary policy on domestic demand.[17] As a result, the deterioration of the current account continued at even greater speed in late 1996 and early 1997.

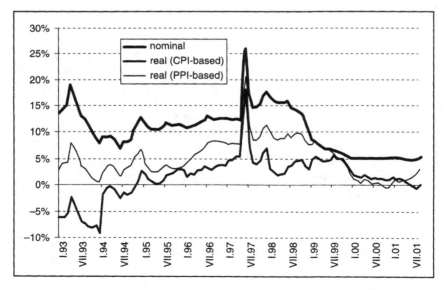

*Figure 4.4*  Money market interest rates (PRIBOR 3M)

*Source*:  Czech National Bank.

The problem was that the monetary restrictions were not accompanied (or perhaps substituted) by a fiscal tightening, that could have been a much more efficient macroeconomic tool for reducing the current account deficit, as fiscal restrictions do not have the negative exchange rate consequences. On the contrary, the fiscal policy became even more expansionary in 1996 than in 1995 (see, for example, Schneider and Krejdl 2000), contributing to further growth of domestic demand and to current account deterioration. The stabilization policy mix must, therefore, be seen as inadequate in 1996.

When growth of output slowed down in 1997, the government fiscal position began to deteriorate which – in the context of worsening external balance – eventually forced the government to implement a package of stabilization measures on 16 April, 1997. This package included fiscal restrictions of CZK 25.5 bn (i.e. 5 per cent of budgeted current spending or 1.5 per cent of GDP), freeze in the state sector real wage bill and a 25 per cent reduction in public sector capital expenditures (for further details on 1997 stabilization packages, see OECD 1998). The package, however, came too late to reverse the negative trends and did not succeed in convincing the investors, whose sentiment was negatively affected by the exchange rate crises in Asian emerging markets. This gave rise to mounting speculations against the crown and resulted in a currency turmoil in May 1997.

In May 1997, the CNB made an attempt to resist the speculative attack by massive interventions, a radical increase in the interest rates (see Figure 4.4) and temporary suspension of the foreign investors' access to domestic credit. Despite these exceptional measures, the CNB was eventually forced to introduce a managed floating of the Czech crown on 26 May, 1997. The exchange rate depreciated immediately by more than 10 per cent (see Figure 4.3). The fiscal restrictions were intensified by another package of stabilization measures (announced on 28 May, 1997), the spending cuts altogether amounting to CZK 45 bn, or 2.7 per cent of GDP. The interest rates were kept rather high throughout the rest of 1997 (see Figure 4.4).

Since 1998, the Czech National Bank has introduced inflation targeting as its new monetary policy regime. This step was motivated by an effort to pin down the inflation expectations and return to the disinflationary path interrupted in the second half of 1997 (see Figure 4.2) due to the CZK depreciation (see Hrnčíř and Šmídková 1998). The interest rates were further increased slightly in early 1998 and maintained at a high level until mid-1998. The exchange rate started to appreciate in 1998, in spite of the ongoing currency crises in other emerging markets. In mid-1998, the nominal effective exchange rate in fact strengthened to its pre-crises level (see Figure 4.3). The combination of high interest rates and a strong CZK meant a substantial tightening of the overall monetary conditions, which had been temporarily relaxed in mid-1997 due to the CZK depreciation (Čihák and Holub 2000). The interest rates started to decline only in the second half of

1998 (see Figure 4.4), and the real monetary conditions did not get substantially looser before the turn of 1999, when the exchange rate depreciated again.

The tight macroeconomic policies, together with an economic slowdown in the EU, contributed to a deepening economic recession during 1998. This factor, coupled with the strong exchange rate and an unexpected decline in food and oil prices, contributed to a substantial slowdown in inflation (see Figure 4.2). As a result, the CNB undershot its targets for net inflation in all three years from 1998 to 2000, even though by a small margin only in the latter case (see Figure 4.5). Although inflation accelerated again in 2000–01, partly due to high oil prices and a return of food prices to their previous levels, the slowdown in inflation can still be considered an important achievement of the macroeconomic stabilization of the Czech economy.

Another important achievement has been the stabilization of the balance of payments. The current account deficit declined to about 2 per cent of GDP in 1998–99 (see Table 4.2). Moreover, the inflow of FDI has sharply accelerated, providing a stable source of financing. Even though the current account deficit again deteriorated in 2000–01 to about 5 per cent of GDP due to a combination of high oil prices and reviving domestic demand, it did

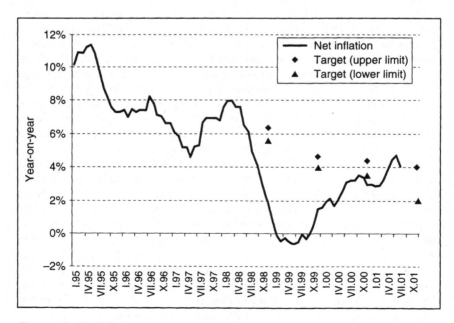

*Figure 4.5*  Net inflation: targets versus reality

*Source*:  Czech National Bank.

not constitute an immediate threat to the macroeconomic balance of the Czech economy. The deficit has been easily financed by the inflow of FDI and, in addition, the negative trend has been reversed since mid-2001 on the back of reduced commodity prices. The rapid FDI inflows have pushed on a gradual exchange rate appreciation – a pressure that was occasionally fought by the CNB interventions. In addition, a special foreign exchange account has been established with the CNB for the government privatization revenues (strengthened much further by an agreement between the CNB and government in January 2002).

To sum up, the initial transition reforms were designed as a consistent package in 1990–91 in which the stabilization policies strongly supported the main reform goals, including those of trade policy (external stabilization, trade liberalization, reorientation of exports). In 1992–93, several imbalances emerged, such as excessive wage growth and real exchange rate appreciation, that later on caused substantial problems for policy makers, but, at that time, the stabilization policies still seemed to be coherent. From 1994 to mid-1997, however, the macroeconomic developments clearly reduced the foreign price competitiveness of Czech companies. The policy of fixed nominal exchange rate, accompanied by sterilized interventions and coupled with a pro-cyclical fiscal policy, proved to be unsustainable. Moreover, it created a favourable environment for short-term capital inflows through the Czech banking system, which further softened the already weak budget constraints in the economy. As a result, the period was characterized by an accelerating domestic demand, persistent inflation and excessive wage growth, which naturally led to a mounting current account deficit. The restrictive steps of monetary policy adopted in mid-1996 were ineffective in reducing the current account deficit in the short-run, as higher interest rates contributed to an exchange rate appreciation that further reduced price competitiveness. Fiscal restrictions would have been more appropriate at that time. Although explicit general government deficits were small, fiscal policy was not restrictive enough to absorb the excess demand. Fiscal stabilization measures were not implemented before April 1997, partly due to the election cycle (1996 was an election year) and partly due to poor understanding of the problem (see Drabek and Schneider 2000). Since mid-1997, a period of successful, but costly macroeconomic stabilization took place. For the central bank, a key question now is how to treat the exchange rate in the new inflation targeting regime. The appreciation of the exchange rate in 1998 apparently contributed to the prolonged economic recession and undershooting of the inflation targets both in 1998 and 1999. In 1999–2000, the CNB seems to have been more concerned with the exchange rate developments, as indicated by occasional foreign exchange interventions and the establishment of special privatization accounts. Nevertheless, it is very difficult for the central bank to resist the appreciation pressures on the CZK, given the massive size of current FDI inflows.

*Table 4.4*  Territorial structure of Czech foreign trade (percentage of total)

| | 1988* | 1993* | 1993 | 1994 | 1995 | 1996 | 1997 | 1998 | 1999 | 2000 |
|---|---|---|---|---|---|---|---|---|---|---|
| **Exports** | | | | | | | | | | |
| Advanced market econ. | 34.2 | 69.1 | 54.3 | 64.5 | 66.0 | 64.5 | 65.6 | 69.5 | 74.6 | 74.8 |
| –of which: EU | 30.2 | 62.9 | 49.4 | 58.8 | 60.5 | 58.5 | 59.9 | 64.2 | 69.2 | 68.5 |
| (Post)communist countries | 57.8 | 20.1 | 37.2 | 29.2 | 28.9 | 30.3 | 29.9 | 26.6 | 22.2 | 21.4 |
| –of which: Slovakia | n.a. | n.a. | 19.7 | 14.6 | 13.9 | 14.2 | 12.9 | 10.6 | 8.3 | 7.7 |
| Other | 8.0 | 10.8 | 8.5 | 6.3 | 5.1 | 5.2 | 4.5 | 3.9 | 3.2 | 3.8 |
| **Imports** | | | | | | | | | | |
| Advanced market econ. | 37.1 | 73.5 | 60.7 | 68.2 | 69.5 | 70.3 | 70.4 | 72.4 | 73.6 | 71.9 |
| –of which: EU | 31.2 | 63.4 | 52.3 | 59.5 | 60.9 | 62.1 | 61.5 | 63.5 | 64.2 | 62.0 |
| (Post)communist countries | 56.9 | 20.4 | 34.3 | 27.4 | 26.4 | 24.9 | 24.8 | 22.9 | 21.6 | 23.3 |
| –of which: Slovakia | n.a. | n.a. | 15.9 | 13.3 | 11.8 | 9.6 | 8.4 | 7.2 | 6.3 | 6.0 |
| Other | 6.0 | 6.1 | 5.0 | 4.4 | 4.1 | 4.8 | 4.8 | 4.7 | 4.8 | 4.8 |

*Source*:  Czech Statistical Office.

*Note*:  * Including trade with Slovakia.

## Trade overview

Here, we shall give an overview of the main foreign trade developments and policies. A more comprehensive summary of trade policy measures in the Czech Republic in the 1990s, however, lies beyond the scope of our analysis (for more detailed analysis of trade policy issues, see, for example, Hrnčíř 2001).

As shown in Table 4.4, almost 60 per cent of Czech exports went to the East in 1988, while less than 35 per cent was directed to advanced market economies. In such a situation, an obvious trade policy goal was to liberalize trade with the West in order to stimulate a territorial reorientation of both imports and exports and, at the same time, avoid excessive trade deficits.

Discriminatory foreign trade restrictions (quotas, tariffs and so on) were substantially reduced during 1991.[18] A range of transitional measures of protection was put in place, though, to prevent a surge in imports while permitting the government to commit itself credibly to a fixed exchange rate that was designed as a nominal anchor in the reform. A temporary flat import surcharge of 20 per cent was introduced in 1990, applied mainly to consumer goods, which was gradually phased out by 1992. Some 20 per cent of merchandise exports was made subject to explicit licensing – as opposed to the former system of implicit quotas and licensing – and on the import side, the licensing covered four categories of goods (crude oil, natural gas, narcotics, arms and ammunition). Access to foreign currency for import purposes was set free but temporary regulations by the State Bank requiring a range of imports to be financed through trade credits were in place between October 1990 and October 1991.[19] A crucial step in further foreign trade liberalization process was the 'association' agreement of Czechoslovakia with the European Union that came into force on 1 March, 1992. This agreement outlined a timetable for a major reduction of mutual trade barriers, the speed of which was generally faster on the EU side than on the Czechoslovak side.[20]

The territorial reorientation of foreign trade proved to be faster than expected. Within two years from the liberalization impulse, some two thirds of Czech exports went to the advanced market economies and their share has been steadily growing since then, notwithstanding the current account crisis of the mid-1990s (Table 4.4).

Another event that affected the territorial structure of trade was the split up of the Czechoslovak Federation. In 1993, a natural trade policy goal was to minimize the adverse impact of the split up on trade between the Czech and Slovak Republics. In order to achieve this, a customs union between the Czech and Slovak Republics was introduced on 1 January, 1993. The customs union agreement established zero tariffs and prohibited quotas in mutual trade and harmonized tariffs and quotas for the trade with other countries. It also contained provisions on public aid, rules of competition, intellectual property rights, legal approximation and so on, all aiming at equalising the legal and economic framework of mutual trade.[21,22] The trade in agricultural products was covered by the customs union agreement, too, but the provisions on common trade policy toward the third countries and on public aid did not apply in this case. The agriculture has proved to be a sensitive area in the subsequent developments, in particular since mid-1996,[23] and there were also other disputes.[24] Nevertheless, the customs union has survived, helping to maintain a considerable degree of foreign trade liberalization between the two republics. Although this was not enough to prevent a steady decline in the relative importance of trade between the Czech and Slovak republics (see Table 4.4), the decline would have probably been even faster and deeper without the customs union.

A further major step in foreign trade liberalization was made in March 1993, when the Central European Free Trade Agreement (CEFTA) entered into force, initially with four participating countries (Czech Republic, Hungary, Poland and Slovakia), that was designed to lead to the establishment of free trade areas after a brief transitional period.

The rising trade and current account deficit, however, challenged the liberal doctrine in foreign trade policy making. As a result, the focus of trade policy clearly started to shift toward more activism by the second half of the 1990s. In 1995, the Czech Export Bank (CEB) was established to provide export financing to Czech companies. While the growing current account deficit was not a primary stimulus for the creation of this institution, it most certainly provided an increased political support and accelerated the process of setting up the CEB. Also, the Export Guarantee and Insurance Corporation (EGAP) has intensified its operations since 1996.

Three anti-import measures were also introduced. Firstly, an anti-dumping law was adopted in June 1997, but the law has not been actively used in practice. Secondly, the shift in trade policy focus was reflected in a recommendation to give preferential treatment to domestic suppliers in public procurement. And finally, an import deposit scheme was introduced as part of the stabilization package in spring 1997. Under the scheme, importers of selected goods (primarily of consumer goods and foodstuffs) were obliged to deposit 20 per cent of the value of imported goods on a non-interest-bearing account for 180 days. As a result, the impact of import deposits on importers' costs could be estimated at roughly 1.5 per cent of the total value of imported goods in April 1997 (given the average interest rate on newly granted credit of about 13.5 per cent at that time). Its impact on costs was, therefore, much smaller than that of an equivalent import surcharge that had been considered as an alternative to the import deposits (but was not used for political reasons). For this reason, the measure could significantly affect only small importers with a limited access to credit, who were thus strongly discriminated by the scheme. Import deposits represented a clear policy slippage that had an adverse impact on the foreign reputation of the Czech Republic. From the macroeconomic point of view, however, they proved to be ineffective and were abolished on 21 August, 1997.

The above anti-import measures notwithstanding, it can be argued that the Czech Republic has relied primarily on export promotion as its major trade policy tool since 1997.[25] Some reference to the need for greater support of trade has been made in the 1997 stabilization packages, the definitive manifestation of this trend was the publication of the Policy Statement by the new Cabinet in August 1998. The government committed itself to create a coherent pro-export policy based on development of the activity of the CzechTrade agency, Export Guarantee and Insurance Corporation and Czech Export Bank. The document envisaged the development of a concept of 'territorial export policy'. Both financial and non-financial measures have been included in the envisaged policy. Financial measures included, most importantly, financing and underwriting of state-supported exports, targeted aid for development, and the system of so-called 'soft loans', that is, long-term government loans on favourable terms. The government also emphasized the need for a greater role of trade diplomacy (including trade representations at its embassies and consulates, and foreign representations of the CzechTrade agency).

In Appendix I, we focus in more detail on the activities of principal trade promotion agencies, namely the Export Guarantee and Insurance Corporation, the Czech Export Bank and CzechTrade. The main conclusions can be summarized in the following way. There has been a significant growth of both financial and non-financial export promotion activities since 1999. Legislative changes and increased budget allocations broadened the scope and scale of the existing institutions and instruments. Despite the sharp increase in financial support, though, its impact on the total volume of exports and the trade balance is still perhaps of the second order (by far more important is, for example, the FDI promotion scheme mentioned on p. 130 and in note 25), and they primarily influence the structure of exports only. Moreover, it is as yet unclear whether increased financial support can be sustained into the future given its growing demand on government funds.

## A model of the Czech current account

### Motivation of the analysis

In this section, we analyse empirically the links between stabilization policies, foreign trade and competitiveness. We put a particular emphasis on studying the sources and consequences of the large current account deficits of 1995–97, which can be justified on three grounds at least. First, the Czech Republic is a small open economy, in which the current account represents a very sensitive indicator of macroeconomic imbalances, revealing possible mismanagement of stabilization policies. Second, the current account developments can be also used to make inferences about the Czech potential GDP growth and equilibrium real exchange rate appreciation, which both represent key uncertainties for macroeconomic policies. Third, the experience with the large external imbalance of 1995–97 greatly affected the subsequent trade and stabilization policy measures (see pp. 124ff). We believe, therefore, that it would be inappropriate to view any of the trade and stabilization policies in the second half of 1990s in isolation from the current account developments. We will try to answer questions such as: 'Was the current deficit of 1995–97 caused primarily by an overvalued exchange rate, overheating of the economy, supply-side bottlenecks or foreign economic slowdown?' 'What was the impact of monetary policy tightening of 1996–97 on the current account?' 'What would have been the appropriate policy mix?'

In 1997, most independent analysts apparently inclined to the overvalued exchange rate hypothesis, and some even predicted the CZK to depreciate above 20 CZK/DEM.[26] However, the balance of opinions has shifted over time. Now many economists, including the CNB's officials,[27] interpret the 1995–97 experience as an economic overheating which showed that the potential GDP growth in the Czech Republic was much slower – due to structural and institutional problems – than the previous forecasts had predicted. The economic recession in 1997–99 is then viewed as a necessary correction that eliminated accumulated imbalances. Nevertheless, there are still many people who put the blame for the negative developments of recent past on allegedly mismanaged stabilization policies (see, e.g., Dyba 1999).

### The data and model specification

We cannot make a serious judgement about this debate without knowing the equilibrium exchange rate of the CZK and potential growth of the Czech economy. This knowledge is critical for our assessment of macroeconomic policies, whether they were successful in performing their job; that is, stabilizing the real variables around their equilibrium levels. Nevertheless, both the potential output and the equilibrium real exchange rate always represent a great source of uncertainty for stabilization policies, which is even more true for a transition economy with its short time series and many structural breaks.

The Czech GDP data cover approximately one business cycle (see Figure 4.6). In this situation, alternative methods for finding the long-run potential may not give robust results. In spite of this, there have been many analyses trying to find out the potential (or equilibrium, non-accelerating-inflation and so on) product of the Czech economy (see, for example, Flek *et al.*, 2001; Hájek and Bezděk 2001; Vašíček and Fukač 2000). The applied methods range from the purely technical to the more theoretical. The former stream has been primarily represented by the well-known Hodrick–Prescott filter, but its use is surely not free of problems, especially at the end of the sample where the filter may follow the actual data too closely.[28] The latter approach can be used, for example, in estimating production functions or in empirical cross-country studies of long-run economic growth (see, for example, Barro 1991). These approaches, however, are also not completely reliable. A simple '$\beta$-convergence' scenario, for instance, would predict an average GDP growth rate of slightly more than 4 per cent for the Czech Republic.[29] Nevertheless, the empirical studies confirm that the

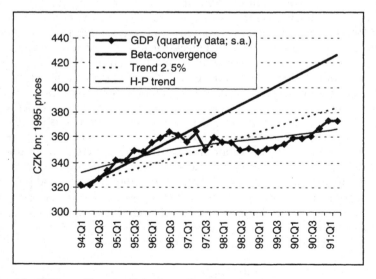

*Figure 4.6*  GDP growth potential: alternative estimates

*Source*:  Czech Statistical Office, own computations.

long-run convergence is conditioned by sound structural and institutional characteristics of the economy. The $\beta$-convergence scenario thus should be interpreted as a benchmark showing what GDP growth rate could have been reached without the existing institutional and structural weaknesses of the Czech economy, rather than as a reliable prediction of the actual growth potential.[30] Similarly, the production function approach is likely to be quite complicated in practice, since there is no reliable proxy for capital stock in the Czech economy. Most recently, the bi-variate Kalman filter was introduced which combines purely technical filtering with some prior knowledge of equilibrium macroeconomic relationships (see, for example, Vašíček and Fukač 2000).[31] While this latter approach seems to be most promising to us, its use would go beyond the scope of this chapter. We shall, therefore, simply compare the GDP data with a Hodrick–Prescott trend, the '$\beta$-convergence' scenario and a 2.5 per cent yearly growth rate trajectory in Figure 4.6.

In fact, all of the alternative methods of analyses (including the Kalman filter methods) lead to the same qualitative conclusions that the Czech economy was overheated in 1996 and below its potential in 1999. But the actual quantification of the potential GDP and output gap differs considerably among individual methods of estimation (see Figure 2.1).

Another source of uncertainty is the equilibrium real exchange rate. Lazarová and Kreidl (1997), for example, argued that the CZK was overvalued by some 10–15 per cent in late 1996. Given the fact that the crown has further appreciated in real terms since then (by 7 per cent in PPI terms and 12 per cent in CPI terms), we might conclude that the exchange rate is probably still fundamentally overvalued. Frait and Komárek (1999), in contrast, found out that the real exchange rate was close to equilibrium until late 1996. This was followed by periods of misalignment (an overvaluation at the turn of 1997, undervaluation after the currency turmoil in mid-1997, and another overvaluation in 1998), but, in general, the real exchange rate had a tendency to return to equilibrium rather quickly. Finally, in the international cross-country comparisons of GDPs and price levels, the Czech Republic lies below the regression line by some 15 per cent (see Figure 4.7). We could conclude from this that the Czech price level is too low, being equivalent to an undervalued real exchange rate. The CZK should thus be appreciating in future as the Czech Republic converges to the EU, which contradicts the view that the crown might be fundamentally overvalued and should depreciate. As with the potential GDP growth, however, the predictions of cross-country studies should be interpreted as a 'normative benchmark' rather than a realistic prediction.

Given these uncertainties, we prefer to avoid using *a priori* estimates in our empirical analysis of the current account developments. Instead, we have decided to choose the opposite strategy; to estimate a fairly standard current account function and to use the estimated coefficients to infer

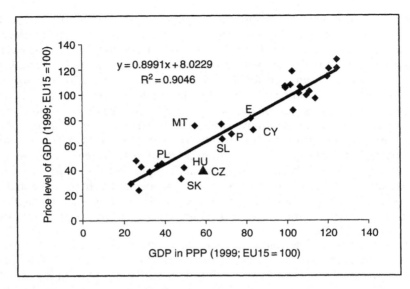

*Figure 4.7*   GDP: price level relationship

*Source*:   Eurostat, Czech Statistical Office.

the size of external constraint on economic growth and the implication of real exchange rate appreciation for the productive potential of the Czech economy.

Let us assume the following functional relationship:

$$CA = \alpha + \beta\frac{Y - Y^*}{Y^*} + \gamma\frac{Y_f - Y_f^*}{Y_f^*} + \delta\frac{ER - ER^*}{ER^*} + \varpi\frac{PC - PC^*}{PC^*} + \varepsilon \qquad (1)$$

where *CA* is the current account balance in per cent of GDP, *Y* domestic real GDP, $Y_f$ foreign real GDP, *ER* real effective exchange rate of the domestic currency, *PC* commodity price index, and the asterisks denote potential/equilibrium levels of these variables.[32] The term $\varepsilon$ denotes a random disturbance, and $\alpha$, $\beta$, $\gamma$ and $\delta$ are constant coefficients. We thus assume that the current account balance depends on the domestic output gap, foreign output gap, the deviation of the real exchange rate from its equilibrium, and the deviations of commodity prices form their long-run equilibrium levels. The intuitive signs of the coefficients are $\beta < 0$, $\gamma > 0$, $\delta < 0$ and $\varpi < 0$.

The key problem in performing an econometric analysis in line with equation (1) is the fact that, as has already been stressed above, the potential output of the Czech Republic and the equilibrium real exchange rate of the CZK are both major unknowns. We thus cannot estimate equation (1) directly.

We can, however, differentiate equation (1) around the equilibrium (where $Y \sim Y^*, Y_f \sim Y_f^*, ER \sim ER^*$ and $PC \sim PC^*$) to get

$$\Delta CA = -\left[ \beta \frac{\Delta Y^*}{Y^*} + \gamma \frac{\Delta Y_f^*}{Y_f^*} + \delta \frac{\Delta ER^*}{ER^*} + \varpi \frac{\Delta PC^*}{PC^*} \right]$$

$$+ \beta \frac{\Delta Y}{Y} + \gamma \frac{\Delta Y_f}{Y_f} + \delta \frac{\Delta ER}{ER} + \varpi \frac{\Delta PC}{PC} + \Delta \varepsilon \qquad (2)$$

where $\Delta$ denotes a time change.

If we now assume that domestic potential output, foreign potential output, equilibrium real exchange rate and equilibrium commodity prices all grow at constant speeds, equation (2) can be rewritten as

$$\Delta CA = \theta + \beta \frac{\Delta Y}{Y} + \gamma \frac{\Delta Y_f}{Y_f} + \delta \frac{\Delta ER}{ER} + \varpi \frac{\Delta PC}{PC} + \Delta \varepsilon;$$

$$\theta \equiv -\left[ \beta \frac{\Delta Y^*}{Y^*} + \gamma \frac{\Delta Y_f^*}{Y_f^*} + \delta \frac{\Delta ER^*}{ER^*} + \varpi \frac{\Delta PC^*}{PC^*} \right] \qquad (3)$$

which is already an equation in observed variables that can be estimated by standard econometric methods. The assumption we used to derive equation (3) is arguably a great simplification, as macroeconomic variables typically exhibit non-linear trends over the longer run. Nevertheless, given the short length of Czech time series, we believe it may be reasonable to proceed this way. We shall estimate equation (3), and then use the stability tests to find out whether there are any structural breaks that might indicate changes in the potential growth rate and/or equilibrium exchange rate appreciation.[33]

From a regression analysis, we get estimates of $\theta, \beta, \gamma$ and $\delta$. We can then write

$$\frac{\Delta Y^*}{Y^*} = -\frac{1}{\tilde{\beta}} \left[ \tilde{\theta} + \tilde{\gamma} \frac{\Delta Y_f^*}{Y_f^*} + \tilde{\delta} \frac{\Delta ER^*}{ER^*} + \tilde{\varpi} \frac{\Delta PC^*}{PC^*} \right] \qquad (4)$$

where symbol '$\sim$' denotes estimated parameters. If we have a reliable estimate of foreign potential growth and equilibrium commodity price growth, equation (4) gives us all combinations of potential GDP growth rates and equilibrium real exchange rate growth rates that are consistent with the estimated structural coefficients of equation (3). This is all we can get from our analysis, since we cannot infer two equilibrium growth rates from just one indicator. But even this limited information may be of great value.

We carried out alternative estimations of equation (3) for Czech quarterly time series. The dependent variable was a change in percentage points of the current account ratio to the GDP over the four subsequent quarters.

We applied the fourth differencing of current account balance to eliminate seasonality. To approximate foreign economic growth, we used a weighted year-on-year growth rate of industrial production in 25 foreign countries that together account for more than 85 per cent of Czech exports. We used industrial production rather than GDP because the time series were available for a larger set of countries. We worked with the 'analytical' real effective exchange rate calculated by the CNB (both on CPI and PPI bases). This indicator is a weighted average of real exchange rates of the CZK against the currencies of the 20 most important trading partners of the Czech Republic (excluding Russia). The variable *PC* was approximated by the CSO's commodity price index. An important data constraint was the quarterly year-on-year growth rates of the Czech GDP, which are available from 1995 only. This reduced the number of observations to 26.

### Estimation results and interpretations

Table 4.5 summarizes the results of our estimations. As we can see, the general characteristics of the models are quite satisfactory. The coefficients of determination are reasonably high for both regressions, considering the fourth differencing of the dependent variable and the relative simplicity of the model. The F-statistics of both regressions are highly significant, and the models have passed the basic stability tests.[34] Based on the Durbin–Watson test, it was not possible to reject the hypothesis of residual independence. The estimated regression coefficients all have the expected signs and plausible values, and are statistically significant at the common probability levels (except for the intercept).

The estimated slope coefficients tell us that:

- if the GDP growth rate exceeds its potential by 1 per cent point, then – *ceteris paribus* – the current account balance deteriorates roughly by 0.45–0.50 per cent of GDP;
- if foreign industrial production grows 1 per cent point faster than its potential, the Czech current account improves roughly by 0.30–0.45 per cent of GDP;
- a real exchange rate appreciation (either PPI- or CPI-based) that exceeds its equilibrium speed by 1 percentage point leads to a deterioration of the current account balance by 0.40 per cent of GDP;

*Table 4.5*  Regression analysis results

| | $\theta$ | $\beta$ | $\gamma$ | $\delta$ *(PPI)* | $\delta$ *(CPI)* | $\varpi$ | $R^2$ *adj.* | *DW* | *P-value* |
|---|---|---|---|---|---|---|---|---|---|
| Time lag | – | 0 | 1 | 1 | 1 | 0 | – | – | – |
| Estimate 1 | 0.62% | −0.49 | 0.31 | −0.39 | – | −0.05 | 0.66 | 1.87 | 0.000 |
| *t-statistics* | (0.67) | (−3.25) | (1.61) | (−4.14) | | (−3.03) | | | |
| Estimate 2 | 0.87% | −0.45 | 0.44 | – | −0.39 | −0.08 | 0.67 | 2.01 | 0.000 |
| *t-statistics* | (0.93) | (−2.99) | (2.38) | | (−4.26) | (−3.91) | | | |

- if the commodity price growth exceeds its long run potential by 10 percentage points, the current account balance deteriorates by 0.55– 0.75 per cent of GDP.

From the estimated slope coefficients, we can directly make some preliminary (and very simplified) judgements concerning the extent of economic 'overheating' and/or real exchange rate 'overvaluation' in 1996. The current account deficit reached almost 8 per cent GDP that year. If we (quite arbitrarily, but in line with the often used 'rules of thumb') assume that a sustainable level of the deficit would be around 5 per cent, the extent of the positive output gap could have been as much as 6 per cent of GDP.[35] [36] In reality, the external imbalance was probably caused by a combination of the two factors. In that case, the 'overheating' was smaller than 6 per cent of GDP, and the overvaluation was also below 7–8 per cent (but we cannot identify the exact combination based on this simple analysis). In a symmetric way we could make judgement also about the period 1998–99, in which the current account deficit declined to roughly 2 per cent of GDP; that is, this time about 3 per cent points *below* the assumed equilibrium.

However, we can be more systematic in expressing these arguments. In particular, we can use the estimated intercept and slope coefficients to quantify the relationship (equation 4). To do this, we also need an estimate of the potential foreign economic growth and the equilibrium commodity price growth. In 1993–2000, the average year-on-year growth rate of foreign industrial production reached 2.9 per cent; we thus used 3 per cent a year as a rough estimate of foreign potential growth. The average commodity price growth reached roughly 2 per cent in the period from 1991 to 2000, which we used as our estimate of its equilibrium speed.

The estimated combinations of potential GDP growth and equilibrium exchange rate appreciation (both PPI- and CPI-based) are presented in Figure 4.8. The horizontal axis shows the sustainable GDP growth rate, while the vertical axis shows the sustainable real exchange rate appreciation. The 'PPI-based equilibrium line' is the estimated equilibrium relationship from 'Estimate 1' using the PPI-based real effective exchange rate, while the 'CPI-based real effective exchange rate. The actual data for the period 1994–2000 are also shown in this figure (the upper mark shows the CPI-based real appreciation, the lower mark shows 'PPI-based' real appreciation), and this period is further divided into two sub-periods of 1994–96 and 1997–2000 (i.e. the time of fast growth and subsequent stabilization, respectively).

We can summarize that:

- the CPI-based equilibrium line lies above the PPI-based line (by about 1.0– 2.0 percentage points). This is quite an intuitive result. The CPI includes prices of non-tradables, and it thus should grow faster than PPI due to deregulations and the Balassa–Samuelson effect. This means that the CPI-based real exchange rate should be appreciating faster than the PPI-based one, which has been indeed the case in the Czech Republic since 1993 (see Figure 4.3);

- if we assume zero PPI-based real exchange rate appreciation in equilibrium, the potential GDP growth can be estimated at about 3 per cent;
- with a PPI-based real appreciation of 3.7 per cent and CPI-based appreciation of 5.2 per cent a year, no 'room is left' for a sustainable GDP growth.

## Policy implications and conclusions

Of course, it is not at our discretion to choose a point on the estimated equilibrium lines. We need some other criterion to find out which point is the 'true' one. For example, we could look at past inflation developments to get some further information about the potential GDP growth. We are not going to proceed this way, however, as this would take us too far away from the subject of this chapter. Instead, we will try to analyse which points (if any) on the lines in Figure 4.8 are reasonable in the sense that they are consistent with a convergence of the Czech Republic to the EU. The 'EU-convergence area' that we have plotted in Figure 4.8 is delimited by two requirements. First, if the Czech GDP is to converge to the EU average, we need to grow by more than 2.5 per cent a year.[37] Second, we should not move further below the regression line in Figure 4.7; that is, the CPI-based real exchange rate appreciation should be at least as fast as the growth differential between the

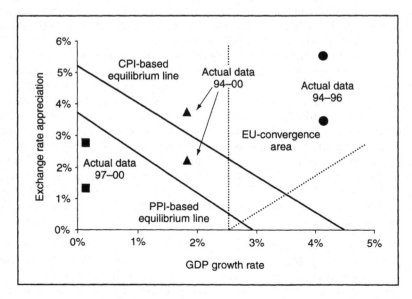

*Figure 4.8*  Estimated potential versus reality

Czech Republic and the EU. We can conclude that the structural characteristics of the Czech economy over the period of 1994–99 closely limited the
set of sustainable economic developments that would have been consistent
with the EU-convergence requirements: a GDP growth of 2.5–3.5 per cent, a
CPI-based real exchange rate appreciation of 1.2–2.2 per cent, and an approximately stable PPI-based real exchange rate (or even a modest depreciation
for the higher growth rates). The speed of GDP convergence in this scenario
is smaller than we could expect based on the conditional convergence literature (see the discussion of Figure 4.6), which may be considered an indirect
proof of institutional weaknesses in the Czech economy.

The reality of 1994–99, however, was much different from the above 'desirable' scenario. In 1994–96, both the GDP growth and speed of real exchange
rate appreciation exceeded the estimated potential. The country was con-
verging to the EU, but the developments were clearly unsustainable. The
current account deficit was caused by a *combination* of economic 'overheating' and excessive real exchange rate appreciation (together with the foreign
economic slowdown of 1996); we thus cannot limit the discussion to only
one of these factors. In *ex-post* terms, we must conclude that the macroeconomic policies failed in stabilizing the economy in 1994–96. The real
exchange rate appreciation, wage growth and domestic demand growth (see
p. 117) were all excessive compared to the economy's potential, hampering
competitiveness and leading to a surge in current account deficit. From an
*ex-ante* point of view, however, the judgement should not be that strict.
First, the data available at that time showed a slower GDP growth for 1995
than the data available at present (the original figure was below 5 per cent,
while the current figure is almost 6 per cent; see, e.g., Dyba 1999). Second,
before the current account deficit built up in 1995–96, many economists had
believed that the potential of the Czech economy was higher than it turned
out to be; after all, the international cross-country studies predict that the
Czech GDP could be growing by at least 4 per cent a year. In our assessment
of the 1994–96 period, we should thus concentrate primarily on the reasons
why the *ex-ante* rational macroeconomic policies were not optimal *ex-post*.
We believe that the blame is to be put primarily on the institutional and
structural weaknesses of the Czech economy that lowered its sustainable
growth rates below the *ex-ante* expected levels (see Drabek and Schneider
2000) and, for an early warning about the institutional weaknesses of the
Czech economic reform, see, e.g., Begg (1992).

In 1997–99, a correction occurred that was characterized by a decline in
GDP and a slowdown in the appreciation of real exchange rate. This correction was associated with the restrictive monetary policy measures introduced
in mid-1996, when the CNB recognized the current account deficit as a
clear signal of macroeconomic imbalances, and with further monetary and
fiscal restrictions necessitated by the currency turmoil of May 1997. Our
analysis suggests that this correction was not excessive. As we can see in

Figure 4.8, the average GDP growth rate reached 1.85 per cent for the period of 1994–2000 as a whole, the CPI-based real appreciation was 3.75 per cent a year, and the PPI-based real appreciation reached 2.25 per cent. These figures are close to the estimated equilibrium lines.[38] In this sense, the correction was just large enough. Nevertheless, it is true that the correction was frustrating from the EU-convergence point of view, as the whole period of 1994–2000 lies outside the EU-convergence area. Purely from this perspective, the desirable pattern of adjustment over 1997–2000 would have been an average GDP growth of about 1.5 per cent accompanied by an average PPI-based real exchange rate depreciation of 3.5–4.5 per cent a year.[39] In simple terms, the desirable adjustment path would have been more 'growth and trade-friendly' than the reality, as it would rely more on a stimulation of exports through a CZK depreciation rather than on a reduction of imports through a sharp restriction of domestic demand.

A question that remains to be answered is whether the described 'optimal' adjustment path was achievable in practice or not. In our opinion, the answer is not clear-cut since policy makers never have the opportunity of fine-tuning the stabilization policies in practice down to such a fine detail. This is even more the case for a transition country such as the Czech Republic. First, there is great uncertainty about the transmission mechanisms of monetary policy due to limited experience and hence short time series, instantaneous structural changes in the economy and so on. Second, there exist (or, at least, existed) institutional bottlenecks contributing to a high inertia of macroeconomic developments (such as the problems in the banking sector in 1998–2000). And finally, the exchange rate developments are influenced by a range of exogenous factors, which make it difficult for the CNB to make active use of the exchange rate channel in its stabilization policy. Examples of these factors include the impact of the emerging market crises of 1997–98, or the high FDI inflow into the Czech Republic since late 1998.

Nevertheless, it was certainly possible to be closer to the 'ideal' adjustment path than in reality. This would, above all, require a different policy mix in 1996. The monetary policy tightening of mid-1996 slowed down the demand growth with a time lag, but at the same time supported the exchange rate appreciation of late 1996 (see p. 124). Its short-run impact on the current account was thus ambiguous – it may even have been negative. The desirable path – that is, a gradual slowdown in demand combined with real exchange rate depreciation – would have required a fiscal tightening instead. But 1996 was an election year, and exactly the opposite happened. In the second half of 1996, the public consumption grew at its greatest speed since 1993, and the structural government budget deficit also reached its (first) peak in 1996 (see, e.g., Drabek and Schneider 2000, Schneider and Krejdl, 2000).[40] The problem thus can be viewed as a failure of coordination between fiscal and monetary policies, which resulted from long recognition and decision lags in the government policy and from political-cycle considerations.

This is not to say, however, that the monetary policy was optimal in *ex-post* terms. Our analysis lends some support to the opinion that the CNB should have tried to prevent the exchange rate from appreciating by faster interest rate cuts, and possibly also by foreign exchange interventions in 1998. At that time, the speculative pressures against the CZK had stopped, fiscal policy was already disciplined, the nominal unit labour cost was declining in the industrial sector, and the domestic demand was falling by more than 5 per cent year-on-year in the first half of 1998. There was thus no reason to let the monetary conditions tighten to their historical maximum (see, e.g., Čihák and Holub, 2000). This, together with a slow-down in foreign demand in late 1998/early 1999, deepened and lengthened the economic recession by reducing the export growth. Once again, however, we stress that it is important to distinguish the *ex-post* judgement from an *ex-ante* decision. In 1998, fears prevailed that the emerging market crises might undermine the stability of the CZK and lead to quick reversals in foreign capital flows. These concerns may serve as an *ex-ante* explanation why the interest rate cuts were slower in 1998 than is now viewed as optimal.[41]

As argued earlier, the CNB's concern about the exchange rate developments seems to have increased somewhat since 1998. In spite of this, it is not realistic to expect that the CZK could start depreciating in the near future, as the rapid inflows of FDI increase the upward pressures on exchange rate. The maximum the CNB can do at present is to try to prevent the CZK from a sharp nominal appreciation. Given the substantial drop in inflation in recent years under the inflation targeting strategy, though, a nominal stability or even a modest strengthening of the CZK has meant a substantially slower real appreciation than was observed in 1992–96. Moreover, the current floating exchange rate regime allows for a flexible, market-driven adjustment of the exchange rate to its equilibrium level if the future developments lead to an excessive current account deficit. The monetary policy thus seems to constitute no problem at present. Nevertheless, there is at least one important risk factor. Many analysts, as well as the CNB officials and foreign institutions[42] warn against increasing structural deficits of the public budgets. As our historical experience shows, a fiscal expansion may be very dangerous in a situation of reviving domestic demand and growing current account deficit (and also decelerating external demand in 2001). If those tendencies thus were to continue as well, there would be a need for more restrictive macroeconomic measures that would prevent the economy from 'overheating' as it happened in 1995–96. If the fiscal policy is not ready to respond flexibly to such a need, which is apparently the case at present, the monetary policy may be forced to do the job. However, this would only mean another sub-optimal policy mix, which is likely to be too costly in terms of slower growth of output and loss of foreign competitiveness.

# Appendix I:  principal export–promotion institutions

In this appendix, we describe in detail the activities of the Export Guarantee and Insurance Corporation, the Czech Export Bank and CzechTrade. Analysis of other policies and agencies lies beyond the scope of this chapter.

## The Export Guarantee and Insurance Corporation

The Export Guarantee and Insurance Corporation (EGAP) was founded in June 1992 as a state owned export credit agency, insuring export credits against political and commercial risks. In addition, EGAP provides also an insurance of domestic receivables.[43]

The Company's main activity is insuring export credits against political and 'non-marketable' commercial risks with state support in the form of a state guarantee for EGAP's obligations arising from this insurance. The second major area is insurance against short-term commercial risks (such as non-payment resulting from a protracted default of a foreign or domestic buyer). This activity is exempt from state support and is offered by EGAP on its own account on a commercial basis. In addition to common products, it also insures pre-export financing of production for export, foreign investment by Czech companies, provides insurance of bank export financing and financing to cover a differential between the domestic and foreign interest rates. Its insurance products are thus similar in scope and structure to those of the foreign export credit agencies in the EU countries.

In the period 1992–2000, EGAP insured export credits to a total amount of CZK 217.5 bn, of which about 55 per cent was state supported country risk. Over this period, it obtained CZK 4 923 mn from the state budget for its country risk insurance funds. Every CZK from the state budget thus enabled insurance of export credits worth CZK 24.75.[44] The EGAP's insurance, however, still concerns some 6 per cent of Czech exports only (see Appendix Table 4AI.1).

*Table 4AI.1*   EGAP: major data

| CZK bn | 1992–95 | 1996 | 1997 | 1998 | 1999 | 2000 |
|---|---|---|---|---|---|---|
| Insurance against country risks | | | | | | |
|    premium | 10.8 | 14.1 | 16.4 | 20.6 | 26.1 | 33.8 |
|    number of new insurance policies | NA | 47 | 76 | 182 | 321 | 267 |
| Insurance against commercial risks | | | | | | |
|    premium | 7.5 | 5.1 | 9.8 | 14.3 | 22.1 | 36.9 |
|    number of new insurance policies | NA | 60 | 60 | 163 | 230 | 171 |
| Overall volume of insurance | 18.3 | 19.2 | 26.2 | 34.9 | 48.2 | 70.7 |
|    as a per cent of Czech exports | 1.0 | 3.2 | 3.6 | 4.1 | 5.2 | 6.3 |

*Source*:   EGAP (annual reports).

The recent development of EGAP's insurance services has been greatly influenced by the above-mentioned change in sentiment among the policy makers toward the export promotion policies. Shortly after coming to power (in November 1998), the Cabinet adopted a resolution on supporting exports[45] allowing, among other things, an expansion and improvement of EGAP services. In particular, an increase in coverage for pure sovereign risk to up to 100 per cent, coverage of the exchange rate risks of paid claims and the introduction of bonuses and discounts to premiums. An amendment to the Act on insuring and financing export with state support[46] was passed by the Parliament in 1999. This amendment, *inter alia*, abolished the condition of a minimum 60 per cent share of Czech deliveries in the export programme as a necessary condition for state-supported insurance and also enabled EGAP to engage in supporting multilateral projects by taking risks corresponding to the share of Czech deliveries. In addition, the new, so-called OECD Consensus rules were implemented,[47] permitting more flexible premium policies that resulted in an average 30 per cent drop in premiums in 1999.

The country risk insurance mainly takes the form of buyer export credit insurance and pre-export finance and has concerned 72 countries since 1992 (most recently 49 countries in 1999, and 46 in 2000). It has an important territorial aspect in supporting the exports to emerging markets, transition economies and so on (Appendix Table 4AI.2), whose relative importance in Czech exports had a tendency to decline during the 1990s.

As far as the insurance against commercial risks is concerned, the share of EU countries in 2000 represented 67.5 per cent of the insured export receivables, 11.1 per cent were non-EU market economies, 18.7 per cent transition economies and 2.7 per cent developing countries. The commodity structure of commercial risks insurance has been biased against the higher value-added commodities with 14.9 per cent machinery, 9.3 per cent industrial consumer goods and 75.8 per cent foodstuffs, raw materials, chemicals and intermediate goods.[48]

*Table 4AI.2* EGAP: country risk insurance by territory (as of Dec. 1999)–(percentages)

| **Belarus** | **11** | **Tunisia** | **2** |
|---|---|---|---|
| Philippines | 8 | Turkey | 14 |
| Yugoslavia | 6 | Turkmenistan | 2 |
| Lithuania | 2 | USA | 2 |
| China | 6 | Other (42 countries below 2 per cent) | 22 |
| Russia | 3 | | |
| Slovakia | 13 | Pre-export financing | 9 |

*Source*: EGAP (annual reports).

Since its foundation, EGAP has thus played an increasingly important role in export promotion policies supporting territorial diversification of Czech exports (especially through the insurance against countries' political risks). Together with the CEB, it creates a necessary financial infrastructure with mainly structural impacts on foreign trade.

### The Czech Export Bank

The Czech Export Bank (CEB) was established as a bank pursuant to Act no. 21/1992 Coll. on Banks. On the basis of a licence to initiate operations issued by the Czech National Bank, the CEB started its operations on 1 July, 1995. Similarly to EGAP, its operations are governed by the Act on insuring and financing export with state support. The CEB is subject to supervision by the Ministry of Finance in the area of state-supported financing and by the Czech National Bank in all matters related to its banking licence.[49]

The basic purpose of the CEB is to promote exports by providing state-supported export financing. As in other such institutions operating in OECD countries (e.g. the US EximBank or EDC Canada), the structure of products is dominated by subsidized medium-term and long-term export credit and bank guarantees. The terms and conditions of its banking and financial services are part of the OECD Consensus. In its capacity as a specialized institution, the CEB complements the existing domestic banking system with its ability to finance export transactions that call for long-term funding sources and that have inherent risks unacceptable to standard commercial banks.

It has been the CEB's founding mission, stated by the government, to use the state-supported financing so that Czech exporters start out on a 'levelled playing field' with their foreign competitors. The CEB thus provides the financial services associated with export on terms and conditions that are more advantageous than the domestic market conditions, in particular with respect to loan maturity and interest rates. This is made possible by state guarantees for the CEB's liabilities, by state budget subsidies to cover the differential between the rates of interest on funds obtained and credits extended, and by the CEB's status as a majority state-owned financial institution. The state guarantee enables the CEB to raise funds on international markets with a sovereign borrower credit rating. The special character of the CEB's financing management consists in the fact that the losses incurred by the CEB from providing financing with state support are subsidized from the state budget. Apart from that, though, the the CEB is a banking institution operating in accordance with rules generally applied to the banking sector.

The years 1999 and 2000 witnessed a significant widening of the CEB's activities following the above-mentioned legislative changes and Cabinet decisions. Total assets increased more than two-fold at the end of 2000, compared to 1998. The same is true about the volume of total outstanding

*Table 4AI.3* CEB: key indicators

| CZK bn | 1996 | 1997 | 1998 | 1999 | 2000 |
|---|---|---|---|---|---|
| Total loans | 3.7 | 7.7 | 8.5 | 12.6 | 18.7 |
| Volume of newly approved loans | 8.7 | 5.4 | 1.7 | 14.8 | 16.5 |
| Exports supported by loans and guarantees | 8.7 | 14.4 | 11.2 | 16.4 | 19.2 |
| as a per cent of Czech exports | 1.5 | 2.0 | 1.3 | 1.8 | 1.7 |
| Subsidies from state budget (CZK mn) | 0.8 | 39.8 | 5.6 | 136.8 | 187.7 |
| Number of employees | 78 | 92 | 97 | 113 | 120 |

*Source*:  CEB (annual reports).

loans, which reached about CZK 19 bn in 2000. While there was a drop in state support in 1998, reflecting the tight budget in the aftermath of the 1997 stabilization packages, the state subsidy grew substantially in 1999–2000 (see Appendix Table 4AI.3). The year 1999 was the first full year of the CEB's operation during which the CEB, owing to preceding legislative steps, provided exporters with a fairly complex set of comprehensive financial services aimed at export support. The CEB has introduced financing of production for exports and also broadened its reach to cover small and medium-sized businesses. Although introduced for the first time in 1999, short-term financing has gained an important position among the activities of the CEB; its current share on the CEB's total loan portfolio shows that short and medium-term loans represent 16 per cent of the total, while 84 per cent are loans with a maturity over 4 years. The most typical product is a short-term direct supplier export revolving loan, usually preceded by the financing of production for export, which also has a revolving nature.

In 2000, the CEB concluded new financing agreements worth CZK 19.2 bn, of which 67 per cent were export loans, 19 per cent financing of production for exports, and 14 per cent guarantees.[50] The state-supported financing represented CZK 15.3 bn, while the rest was financing on market terms. The territorial structure changes from year to year and is influenced by important business cases in progress.[51] The commodity structure reflects CEB's focus on long- and medium-term financing of loans provided for exports of investment units for the energy sector (42 per cent of the state supported financing in 2000), transportation equipment (21 per cent) and machinery (6 per cent).

Based on Appendix Table 4AI.3, it can be argued that the CEB has not had a decisive influence on the total volume of Czech exports so far. Its main influence probably has been concentrated on only some commodity groups and territories, and has therefore been rather structural in nature. Nevertheless, with changing overall sentiment towards export promotion, we can see a significant increase in its activities and in the size of state subsidies in the first two years of the 1990s. It remains to be seen, however, whether this change can be sustained in the following years as the government faces growing difficulties to control mounting fiscal imbalances.

### Trade promotion organization

The Czech Trade Promotion Agency – Czech Trade – was established in 1997 as a government agency of the Ministry of Industry and Trade following the approval of the 1997 stabilization packages. Its mission has been defined as 'Czech trade promotion and strengthening of the trade representation in key territories with activities that would be comparable with the standards of similar trade promoting organisations abroad'. This non-financial export support includes a wide range of activities, such as advisory and consulting services to exporters and importers, education, promotion activities abroad, direct assistance on foreign markets, identification of business partners and suppliers both from the Czech Republic and abroad, marketing studies, consultations and information dissemination regarding public procurement projects and tenders abroad and so on, with a particular emphasis on small and medium-sized enterprises.

In 1999, the activities of CzechTrade were formalized by instituting a special government programme 'Export Promotion' to be administered by the agency.[52] The programme is targeted at the small and medium-sized companies with the number of employees not exceeding 250. In 1999, some 375 companies participated in the programme and the total amount of funds allocated has been CZK 35 mn. In 2000, CzechTrade approved 624 applications worth CZK 82 mn, of which CZK 64 mn was actually drawn by companies during the year. While this amount of funds committed to the marketing form of trade promotion has been rather trivial, the creation of CzechTrade has perhaps had an important signalling effect, showing the increased willingness of authorities to engage in all sorts of active trade policies.

## Appendix II:  a model of trade balance for Poland

We were interested whether the methodology applied to the study of the Czech current account would also give reasonable results for other countries. We have chosen Poland, the developments of which resemble in some respects the Czech situation of 1996–97.

*Table 4AII.1*  Regression analysis results for Poland

|  | $\theta$ | $\beta$ | $\gamma$ | $\delta(CPI)$ | $R^2 adj.$ | $DW$ | $P$-value |
|---|---|---|---|---|---|---|---|
| Time-lag | – | 0 | 1 | 0 | – | – | – |
| Estimate 1 | 0.41 | −0.60 | 0.75 | −0.19 | 0.87 | 2.42 | 0.000 |
| *t-statistics* | (0.75) | (−5.87) | (7.33) | (5.89) | | | |

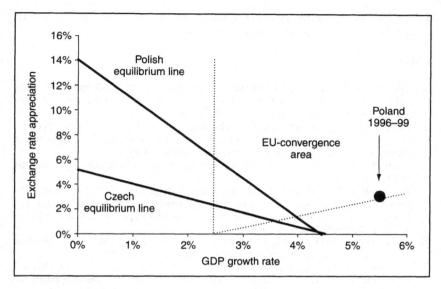

*Figure 4AII.1* Estimated potential for Poland

Unfortunately, we were even more restricted by data problems in the Polish case than in the Czech study. The quarterly GDP growth rates were only available starting from 1996. The available quarterly current account series were too short, so we had to estimate the model for the trade balance only. The results were:

Figure 4AII.1 compares the estimated equilibrium line for Poland with that for the Czech Republic (for CPI-based real exchange rates), and the average Polish data for 1996–99. The equilibrium line for Poland lies above the Czech one, which means a higher estimated growth potential for the Polish economy compared to the Czech economy (but even the Polish estimate is below its $\beta$-convergence forecast of 5–6 per cent). When we compare the actual data with the estimated potential, the Polish economy appeared to be 'overheated' in 1999 (even though to a lesser extent than the Czech economy was in 1996). We believe that this has been confirmed by the subsequent developments. An adjustment was, therefore, needed. However, the 'Polish adjustment' appears to have been somewhat facilitated by the reliance on floating exchange rate policy and strong growth of FDI. Nevertheless, it seems that a wrong policy mix was applied which led to 'hard lending' in the same way as it happened in the Czech Republic in 1997–99.

Finally, we must note that the estimated model for Poland exhibited signs of instability over time. The results have therefore to be interpreted with caution. It goes beyond the scope of this chapter to investigate whether this instability is due to structural changes in the economy, wrong model specification or other econometric problems.

## Notes

1   Headline inflation averaged less than 2 per cent between 1980 and 1989 while hidden inflation was estimated at about 2.5 per cent per annum over the same period. See Drabek *et al.*, 1993; and OECD, 1991.

2   The gross foreign debt of Czechoslovakia amounted to USD 7.9 bn (14.9 per cent of GDP) in 1989, compared to 58.8 per cent in Poland and 71.3 per cent in Hungary. Overall, Czechoslovakia was a net foreign creditor.

3   Council for Mutual Economic Assistance (CMEA), socialist countries' trading block, was dissolved in 1990.

4   The value-added tax system was introduced in January 1993, replacing the turnover tax. At the same time, a personal income tax was introduced and health and social security charges started to be levied as a fixed percentage of employees 'earnings and employers' wage bills.

5   For a detailed description of the macroeconomic developments in the first four years of economic transition, see, for example Hájek *et al.* (1995).

6   The new exchange rate was fixed at 28CZK per USD in December 1990. This was then about the middle ground between the parallel market rate and the estimated average cost of earning one dollar of foreign exchange, and it was roughly four times the PPP rate (see OECD 1991). A good overview of the exchange rate policy in the early years of transition is, for example, provided in Hrnčíř (1993).

7   Later on, the basket was simplified to a weighted average of the DEM (65 per cent) and USD (35 per cent).

8   The price developments in the early stage of transition were analysed, for example, by Tůma (1993).

9   The monetary restrictions also had their negative side effects. In addition to the economic recession, which was mentioned above, there was a surge in the involuntary inter-enterprise arrears (i.e. debt) (see Tůma 1992) that later on caused problems in the enterprise restructuring.

10   In 1991, however, the average real exchange rate already appreciated in terms of the Producer Price Index (see Table 4.3).

11   Dyba and Švejnar (1995), for example, stress that the real wages were still 20 per cent below their 1989 level at the end of 1993.

12   A detailed description of the monetary policy developments during the early phase of transition is presented, for example, in Bulíř (1993) or Guba and Skolková (1993).

13   For an earlier debate of the currency convertibility and its implications on the exchange rate policies, see, for example Dědek (1995).

14   *Nota bene*, the law fully abolished the foreign exchange supply obligation on the domestic companies, allowed opening of foreign exchange accounts with the domestic banks and accepting credits from non-residents, further liberalized FDIs and purchases of real estate abroad and so on. Some activities still remained subject to foreign exchange permission by the CNB (e.g. opening accounts with foreign banks, granting credits to non-residents, issuing foreign securities in the domestic market, derivative operations, etc.). In general, the inflow of capital was liberalized more than its outflow. Most of the remaining restrictions were lifted in 1999–2001, and only a few remain at present (such as restrictions on purchases of Czech real estate by non-residents).

15  This was only partly compensated by a positive effect of the ongoing trade liberalization.

16  Real wage growth exceeded productivity growth by a wide margin throughout the period of 1994–97, which would have led to an acceleration in inflation had this pressure not been offset by a deflationary pressure from import prices. Cincibuch and Vávra (2000) provide an interesting decomposition of the real exchange rate's appreciation since 1993, and conclude that a substantial part of it is attributable to an excessive wage growth.

17  On p.p. 131ff, we estimate the effects of a real effective exchange rate on the current account balance (and thus net exports) with a lag of one quarter. The transmission lag of the effect of monetary policy on domestic demand and inflation is typically thought to range from two to six quarters.

18  The trade policy measures and foreign trade developments during 1991 were described, for example, in Brabec (1992).

19  See Hrnčíř (2001) for further details on trade policy measures.

20  Drábek (1994), however, disputes the view that the 'association agreement' was so asymmetrically favourable for Czechoslovakia in practice.

21  It should be noted that the customs union between the Czech Republic and Slovakia is different in its origin and spirit to most traditional customs unions. Unlike in the traditional case, its purpose was not to increase the trade liberalization between countries, but to *sustain* as much of the existing degree of mutual trade liberalization as possible. A traditional customs union also usually starts with a larger set of exceptions from general rules and then tries to move further to eliminate some of these exceptions, while the Czecho – Slovak customs union moved in the opposite direction – from a large degree of liberalization to more 'backsliding'.

22  An earlier detailed discussion of the customs union between the Czech Republic and Slovakia can be found, for example in Soukup (1995).

23  In late 1996, the Slovak Republic announced its intention to introduce quotas (two of them equivalent to complete prohibition of imports) on eight selected agricultural products imported from the Czech Republic. The Slovak Republic, however, withdrew the proposed measures after mutual negotiations. In April, however, the Slovak Republic introduced quotas on Czech imports on non-alcoholic beverages and considered the possibility of completely excluding agricultural products from the customs union. In June 1997, the Slovak government set an import quota on Czech beer. Another example is an introduction of import quotas on Czech pork and live pigs in April 1999 (abolished in November 1999).

24  For example, the Slovak Republic introduced an import surcharge several times (a 10 per cent surcharge introduced in March 1994, reduced to 7.5 per cent in mid-1996 and abolished at the end of 1996; a 7 per cent surcharge replacing import deposits introduced in July 1997, abolished in October 1998 and reintroduced in June 1999).

25  It should be noted in this respect that these pro-export policies constitute a part of the wider picture of growing state interventionism under the new (social-democratic) cabinet. Of the policies with a potential indirect and structural impact on trade, one should mention the enactment of an FDI incentive scheme (administered by a specialised government agency CzechInvest), 'revitalization' programme aimed at restructuring ailing large industrial enterprises, providing direct subsidies to industrial enterprises, and so on.

26  See, for example, MF Dnes, September 18, 1997, page 14.

27  For the CNB's official views, see ČNB (1999).

28  Short time series may be a problem when using the Hodrick–Prescott filter. Nevertheless, this method also has some general shortcomings that apply even for countries with sufficient data – see, for example, St-Amant and Norden (1997). One of the problems is the relative unreliability of this filter at the end of the sample. To partly reduce this problem when using the Hodrick–Prescott filter, Hájek and Bezděk (2000) included the CNB forecast of real GDP for 2000 and 2001 into the data set.

29  Fisher *et al.* (1998) calculated the expected potential GDP growth rate of the Czech economy at 4.2–4.6 per cent.

30  Campos (1999) shows that the transition countries do not fit well into the results of international cross-country studies of long-run economic growth.

31  The Kalman filter has been also used by the CNB's model forecasting team as one of the alternative methods. Unfortunately, no reference to a published work can be provided here.

32  By specifying the functional relationship in this way, we admittedly ignore the dynamic impact of increasing foreign debt on the current account due to interest payments. A more complete, dynamic analysis would need to include these elements. However, we believe that the qualitative interpretations of our simple model would not significantly change.

33  Also note that equation (4.3) is a fairly standard specification in the current account analysis. We make the underlying assumptions explicit, instead of keeping them implicit only as in many other papers. Moreover, we illustrate in Appendix II that our methodology yields reasonable results not only for the Czech Republic but also for Poland, which has experienced similar monetary problems to the Czech Republic.

34  The recursive coefficient test, however, indicated that the coefficient on foreign industrial production might be declining over time. This may be caused by changes in the territorial (and perhaps also commodity) structure of foreign trade. It is too early, though, to draw strong conclusions in this respect.

35  To achieve a reduction in the current account deficit by 3 percentage points, the GDP would have had to be lower by 6 percentage points, given the estimated elasticity of roughly 1/2.

36  Obtained by taking 3 percentage points of the current account imbalance and dividing by the estimated coefficient of 0.4.

37  There is, of course, no guarantee that the Czech Republic must indeed converge to the EU. But if it does not, we have to interpret this as a clear sign of institutional and structural failures that prevent the 'conditional convergence' from materializing.

38  In fact, they are still slightly above the estimated equilibrium lines. This difference, however, is well within plausible statistical error. Moreover, we cannot exclude the possibility that the Czech Republic was slightly below its potential at the end of 1993 (when it was just emerging from the transition recession and a temporary slowdown caused by the split-up of Czechoslovak federation). We thus would not argue that the correction of 1997–2000 was insufficient. In fact, we tend to believe that the Czech economy was still slightly *below* its potential in 2000.

39  This scenario seems to be broadly in line with what most analysts forecast in mid-1997 (see, e.g., Janáček *et al.*, 1998). Moreover, it would also probably have been more desirable from the perspective of achieving CNB inflation targets, as both a CZK depreciation and slightly positive GDP growth would have pushed on higher inflation in 1998–99 compared to its actual level.

40 This was not because of the elections but partly due to the fact that the GDP growth was overestimated in the 1996 budget draft.

41 For this interpretation, see IMF: 'Czech Republic: Staff Report for the 2000 Article IV Consultation', August 2000.

42 See, for example Bezděk and Matalík (2000), Minutes of the (CNB's) Board Meeting on 26 October 2000, or the IMF Czech Republic, Staff Visit Concluding Statement, October 2000 (both available at http://www.cnb.cz).

43 EGAP is a 100 per cent state owned joint stock company with equity of CZK 1300 million. The state's shareholder rights are jointly exercised by the Ministry of Finance (40 per cent), Ministry of Industry and Trade (36 per cent), Ministry of Agriculture (12 per cent) and Ministry of Foreign Affairs (12 per cent).

44 The multiplier effect of budgetary funds on trade is even higher since the total value of export contracts is always higher than the value of the insured part. EGAP covers at most 85 per cent of the total contract value for long-term credits, for letters of credit the coverage is typically 5–10 per cent, and so on.

45 Cabinet resolution no. 747 of 1998.

46 Act no. 58/1995 Coll.

47 The rules falling within the framework of the Arrangement on Guidelines for Officially Supported Export Credits. See http://www.oecd.org/news_and_events/release/nw96-11a.htm for details.

48 In 2000, the countries with the highest share in newly provided country risk insurance were: Turkey (25.1 per cent), Russia (12.9 per cent), Belarus (12.5 per cent), Slovak Republic (9.3 per cent), Iran (6.3 per cent), Bulgaria (4.1 per cent) and USA (4.1 per cent). The remaining countries accounted altogether for 8.4 per cent only. The share of pre-export financing was 15.3 per cent.

49 CEB is a joint stock company with registered capital of CZK 1.65 billion (increased from CZK 1.5 billion in 2000). The founders and shareholders of the Bank are EGAP with a 33 per cent share and the Czech state with a 67 per cent share. As in the case of EGAP, the state's shareholder rights are jointly exercised by the Ministry of Finance (35 per cent), Ministry of Industry and Trade (20 per cent), Ministry of Agriculture (4 per cent) and Ministry of Foreign Affairs (8 per cent).

50 Guarantees are usually granted only for a certain portion of an export contract (mostly for 5–10 per cent, based on the type of the guarantee). The volume of guarantee contracts is not therefore important as such since the guarantee contracts only provide support to a fraction of the amount of exports, which is many times greater.

51 The USA, Turkey, CIS and Slovakia were the most important territories of state-supported export financing in 2000. Similarly, Turkey and CIS were pivotal in 1999. In 1998, the dominant territories were Phillipines, Iran, Tunisia and Slovakia.

52 Under the Czech budgetary rules, the state support can only be provided to enterprises within a framework of a government-approved programme.

# References

Barro, R.J. (1991) 'Economic growth in a cross section of countries', *Quarterly Journal of Economics*, vol. 106, no. 2, May, pp. 407–43.

Begg, D. (1992) 'Economic reform in Czechoslovakia: Should we believe in Santa Klaus?', *Politická ekonomie*, vol. 40, no. 2, pp. 156–98.

Bezděk, V. and I. Matalík (2000) 'Riziko na straně vlády', *Ekonom*, no. 39/2000, pp. 24–6.

Brabec, J. (1992) 'Československý zahraniční obchod v prvém roce ekonomické reformy', *Finance a úvěr*, vol. 42, no. 4, pp. 176–86.

Bulíř, A. (1993) 'Československá monetární politika po roce 1989' ('Czechoslovak monetary policy after 1989'), *Finance a úvěr*, vol. 43, no. 5, pp. 181–200.

Campos, N.F. (1999) 'Back to the future: the growth prospects of transition economies reconsidered', Prague: CERGE, WP no. 146.

Čihák, M. (1997) 'Ohlédnutí za fluktuačním pásmem koruny', *Finance a úvěr*, vol. 47, no. 10, pp. 608–18.

Čihák, M. and T. Holub (2000) 'Indexy měnových podmínek', *Finance a úvěr*, vol. 50, no. 12, pp. 654–72.

Cincibuch, M. and D. Vávra (2000) 'Na cestě k EMU: Potřebujeme fixní měnový kurz?', *Finance a úvěr*, vol. 50, no. 6, pp. 361–84.

ČNB (1999) *Cílování inflace v České republice*, Prague: CNB (http//:www.cnb.cz).

Dědek, O. (1995) 'Currency convertibility and exchange rate policies in the Czech Republic', *Politická ekonomie*, vol. 43, no. 6, pp. 723–49.

Drabek, Z. (1994) 'Výzva k úpravě podmínek Evropské dohody', *Politická ekonomie*, vol. 42, no. 3, pp. 369–82.

Drabek, Z., K. Janáček and Z. Tůma (1993) 'Inflation in Czechoslovakia', Washington, DC: World Bank Working Paper WPS 1135.

Drabek, Z. and O. Schneider (2000) 'Size of public sector, contingent liabilities and structural and cyclical deficits in the Czech Republic, 1993–1999', *Post-Soviet Geography and Economics*, vol. 41, no. 5, p. 311–41.

Dyba, K. (1999) 'Macroeconomic policy and economic growth during the transition: the case of the Czech Republic in the 1990s', *Eastern European Economics*, vol. 37, no. 4, July–August, pp. 5–21.

Dyba, K. and J. Švejnar (1995) 'A comparative view of economic developments in the Czech Republic', in J. Švejnar (ed.), *The Czech Republic and Economic Transition in Eastern Europe*, San Diego, CA, Academic Press.

Fischer, S., R. Sahay and C. Vegh (1998) *From Transition to Market: Evidence and Growth Prospects*, Washington, DC: IMF, WP/98/52.

Flek, *et al.* (2001) 'Výkonnost a struktura nabídkové strany', Prague, CNB, VP 27–01.

Frait, J. (1996) 'Autonomie monetární politiky a monetární přístup k platební bilanci (aplikace na ČR v letech 1992–1995)', *Finance a úvěr*, vol. 46, no.5, pp.266–80.

Frait, J, and L. Komárek (1999) 'Dlouhodobý rovnovážný reálný měnový kurz koruny a jeho determinanty', Prague: ČNB, VP9–99.

Guba, M. and M. Skolková (1993) 'Transformance čs. ekonomiky a vývoj měnové politiky' ('Transformation of the Czechoslovak Economy and Monetary Policy Development'), *Finance a úvěr*, vol. 43, no. 11, pp. 493–521.

Hájek, M., *et al.* (1995) 'Makroekonomická analýza české ekonomiky (1990–1994)', *Politická ekonomie*, vol. 43, no. 1, pp. 175–86.

Hájek, M. and V. Bezděk (2000) 'Odhad potenciálního produktu a produkční mezery v ČR', Prague: CNB, mimeo, November.

Hrnčíř, M. (1993) 'Exchange rate and the transition: the case of the Czech Republic', *Politická ekonomie*, vol. 41, no. 4, pp. 439–60.

Hrnčíř, M. and K. Šmídková (1998) 'The Czech Approach to Inflation Targeting', Prague: CNB, Workshop on Inflation Targeting, proceedings of a conference held on September 14–15.

Hrnčíř, M. (2001) 'Trade Policy in the Czech Republic in the Decade of Transition', (unpublished), Paper prepared for the Phare ACE Project on Stability of Trade Policy, No. P97–8129-R, directed by Zdenek Drabek.

Janáček, K., M. Čihák, M. Frýdmanová, T. Holub and E. Zamrazilová (1998) 'Czech Economy in 1998: Risks and Challenges', *Prague Economic Papers*, no. 2, pp. 99–139.

Jonáš, J. (1996) 'Měnová politika a měnový kurz' ('Monetary policy and the exchange rate'), *Finance a úvěr*, vol. 46, no. 1, pp. 11–25.

Klacek, J. and M. Hrnčíř (1994) 'Macroeconomic Policies: Stabilization and Transition in Former Czechoslovakia and in the Czech Republic', Warsaw: Centre for Social and Economic Research, March.

Lazarová, Š. and V. Kreidl (1997) 'Rovnovážný měnový kurz', Prague: IE CNB, VP no. 75.

OECD (1991) *OECD Economic Survey: Czech and Slovak Federal Republic*, Paris: OECD.

OECD (1994) *OECD Economic Survey: The Czech and Slovak Republics*, Paris: OECD.

OECD (1998) *OECD Economic Survey: Czech Republic*, Paris: OECD

Schneider, O. and A. Krejdl (2000) 'Strukturálni schodky veřejných rozpočtú v ČR', *Finance a úvér*, vol 50, no. 3, pp. 160–74.

Šmídková, K. (1996) 'Účinnost měnové politiky a poptávka po penězích', *Finance a úvěr*, vol. 46, no. 8, pp. 471–80.

Soukup, P. (1995) 'Celní unie mezi ČR a SR', *Politická ekonomie*, vol. 43, no. 2, pp. 175–86.

St-Amant, P. and S. van Norden (1997) *Measurement of the Output Gap: A Discussion of Recent Research at the Bank of Canada*, Ottawa, Ontario: Bank of Canada.

Tomšík, V. (2001) 'Regresní analýza funkcí zahraničniho obchodu ČR v letech 1993–1998', *Finance a úvěr*, vol. 51, no. 1, pp. 46–58.

Tůma, Z. (1992) 'Monetary and Fiscal Policies during Transition', Prague: Charles University, CERGE, January.

Tůma, Z. (1993) 'Liberalizace cen a vývoj cenové hladiny' ('Liberalization of Prices and Price Level Development'), *Politická ekonomie*, vol. 41, no. 2, pp. 187–97.

Vašíček, O. and M. Fukač (2000) 'Quantitative Analysis of Non-Accelerating Rate of Unemployment and Non-Accelerating Inflation Product', Brno: Faculty of Economics and Administration, Masaryk University, mimeo, November.

# Part III
# Trade Discipline in the Accession Countries of Central Europe

# 5
# Poland: Stability of Trade Policy and its Determinants

*Jan Michalek*

## Introduction: macroeconomic stabilization and exchange rate policy

The overwhelming victory of Solidarity in the parliamentary elections of June 1989 marked the end of the old regime in Poland. The new government presented an outline for economic reform (known as the Balcerowicz Plan) in December of that same year. The main economic objectives were to curb inflation, restore general market equilibrium, and stimulate competition with the intention of leading the system towards a market economy. The restoration of domestic market equilibrium was achieved mainly by liberalizing prices from administrative controls. This was accompanied by an anti-inflationary policy which consisted of a relatively restrictive mix of monetary and fiscal policies as well as wage controls.

The opening up of the Polish economy was an important element of the Balcerowicz Plan. Trade liberalization was seen as a necessary condition in the restoration of the market equilibrium and the promotion of competition in the domestic market, which was dominated by large state-owned monopolies. Moreover, a stable exchange rate (for at least three months) was regarded as a nominal anti-inflationary 'anchor', stabilizing prices and diminishing inflationary expectations. Thus, the Balcerowicz plan was a standard heterodox macroeconomic stabilization programme, elaborated in consultation with the IMF. The main innovation was that such a programme was to be applied for the first time to a non-market economy.

Compared with the situation in Czechoslovakia and Hungary, there were some important features of the Polish economy in the late 1980s which had an impact on the evolution of trade policy in the 1990s.

First, Poland was an example of the economy of shortages at its extreme. The shortages were common, and the black market was flourishing. Until the end of 1989, credit was rationed and the real interest rate was strongly negative.[1] In 1989, the fiscal deficit achieved 7.5 per cent GDP and the inflation rate, after price liberalization, rose to 640 per cent.[2] Consequently, no room was left for a gradualist approach to reforms. A radical macroeconomic stabilization programme was a must at that time.

Second, agriculture and farmers played an important role in the economy. In the late 1980s, private farms occupied 77 per cent of Polish farmland. The relevant figures for the Czech Republic and Hungary were 1 per cent and 6 per cent, respectively. On the other hand, even in the mid-1990s, as much as 27 per cent of the Polish population was employed in the agricultural sector, against 6.3 per cent and 8.0 per cent, respectively, for the Czech Republic and Hungary.[3] It should also be mentioned that farmers provided strong political support for the Solidarity movement in the early 1990s, and the Polish Peasants Party (PSL) consistently played an important role on the political scene. In consequence, every government in Poland during the 1990s had to be quite sensitive to the political claims of these farmers – especially given the higher than average employment and a much lower than average per capita income in rural areas. Therefore, the agricultural lobbies' calls for agricultural interventionism and protectionism could hardly be ignored.

Third, Poland's status in GATT at the beginning of the transformation process was quite different. Czechoslovakia became a GATT contracting party back in 1947 and remained an 'inactive' member in the whole postwar period, even though its economy was a non-market one. Its tariffs, of little economic importance, became bound and quite low in the course of several GATT rounds of multilateral trade negotiations. Poland, when applying for GATT membership in 1958, did not want to make significant change to its non-market economic system. In the course of lengthy negotiations, both sides recognized that tariffs were not an important trade policy instrument. In consequence, a unique reciprocity formula was elaborated in the Polish Protocol of accession to GATT in 1967. In exchange for the most favoured nation (MFN) status among GATT members, Poland committed itself to increase the value of its imports from the territory of GATT contracting parties by at least 7 per cent annually.[4] This untypical commitment became obsolete and impossible for Poland to fulfill as early as the mid-1970s. But, formally, nothing was changed until the creation of the World Trade Organization (WTO) in 1994. Finally, Hungary became a GATT member in 1973. The country offered standard tariff concessions, which, with some reluctance, were accepted by other contracting parties. So, in fact, only Poland, among the three analysed countries, had 'unbound' duties within the framework of GATT at the beginning of transformation. The leeway in trade policy was therefore much larger in Poland than in the other two countries.

As already mentioned, introducing an internal convertibility of the zloty (Polish currency) was an important element of the Balcerowicz Plan. When Solidarity assumed power, the dollar bought at the free market was 40 times as high as the official rate, but the repeated devaluations in the second half of 1989 narrowed the spread significantly.[5] On 1 January, 1990 the official rate for all commercial transactions was set at 9,5000 zloty/dollar. In determining the new rate the government took into account, among others, the following factors: (i) the effects of the withdrawal of export subsidies and other export preferences, and (ii) the high probability that for quite some time Polish prices would rise faster than those of the country's trading partners. For the latter reason, the zloty was deliberately undervalued. Though the devaluation was expected to improve the trade balance, the government forecast that the current account deficit would widen from $2 billion in 1989 to $3 billion (7 per cent of GDP) in 1990.[6]

The transformation process also brought about important changes in Poland's trade policy. The first attempts to liberalize trade were undertaken by the communist regime towards the end of 1987. The criterion for obtaining a permit to engage in foreign trade was then eased and the list of goods which required this permit was substantially reduced. But this liberalization had little impact on foreign trade because most of the foreign exchange (60 per cent in 1989) was still allocated in accordance with annual plan directives. By the end of that year, 21.5 per cent of total exports to market economy countries was still unprofitable and had to be subsidized.

A genuine trade liberalization only began in January 1990, when the zloty became internally convertible and almost all domestic prices were released from administrative control. Thus, subsidies in the form of non-tariff measures were eliminated. The new customs code, introduced in January 1990, was compatible with international norms. The tariff description, the rules of customs valuation, and the anti-dumping procedures were principally in line with the GATT articles. The tariff structure was not fully adapted to market economy requirements, and was somewhat arbitrary. The average tariff level was 8.9 per cent *ad valorem* at the beginning of 1990.

The first six months of 1990 brought important changes in the Polish economy. Inflation was reduced significantly: the monthly retail price index decreased from 180 in January to 103.6 in June. Socialized sector output dropped by 17 per cent in the first two months and real wages fell dramatically. The removal of quantitative import restrictions released demand, and imports surged in January and February. But, from March, imports were drastically reduced in the wake of undervaluation of the zloty and a sharp decrease in aggregate demand. A surprising current account surplus was registered. In order to reduce import charges, customs duty collection was suspended in July–August 1990 for some two thirds of dutiable items, lowering the trade weighted protection from 8.9 per cent to 4 per cent.[7] This, however, did not apply to agricultural goods and other products 'sensitive' to foreign competition, as well as some industrial consumer goods. The government regarded the suspensions as a temporary solution only.

In the middle of 1990, a Commission for Tariff Revision was set up. Its task was to draft a new tariff schedule, adjusted to the market economy system. Changes were proposed in view of the planned negotiations with GATT, and, in connection with negotiations to create a free trade area with the European Communities, EFTA, and CEFTA.

By the end of 1990, Polish prices almost reached the world level; that is, the fixed exchange rate was beginning to hold. Foreign competition squeezed the profits of Polish enterprises. Revenue from corporate income tax declined and – with no matching improvements on the fiscal expenditure front – the budget went into deficit. The situation became more difficult in the course of 1991. In January of that year, the retail price index jumped by 12.7 per cent as a result of drastic increases in administered rates of housing rents, public utility charges and transport fares. In subsequent months, the rate of inflation was gradually reduced to less than 1 per cent per month. But the fiscal deficit fuelled cost-pushed inflation, aggravating profit squeeze. On 17 May, 1991 the zloty was devalued by 17 per cent in relation to the dollar in order to compensate for domestic inflation. The Polish currency was de-linked from the dollar and linked to a basket of currencies (45 per cent US dollar, 35 per cent Deutsche Mark, 10 per cent pound sterling, 5 per cent Swiss and French francs). But it was only a short-term remedy. In October 1991, a crawling peg with a 1.8 per cent monthly rate of devaluation replaced the fixed exchange regime.

On 1 August, 1991, a new tariff schedule came into force. It adopted the Combined Nomenclature (CN) applied by the European Community. The tariff nomenclature covered 10,000 tariff items and the level of duties was substantially raised. Average nominal (unweighted) customs rates calculated on the basis of MFN were raised from 11.65 per cent to 17.02 per cent. But when duty suspensions were also taken into consideration, the average level was raised from 5.82 per cent to 16.83 per cent (the respective figures for weighted duties with suspensions were 5.49 per cent and 14.27 per cent).[8] This was feasible because Polish tariffs were unbound in GATT.

'The increase was motivated by the need to increase fiscal revenues, and by the desire to afford a degree of protection to Polish producers competing with imports. Among products that were granted high protection were particularly "sensitive" agricultural goods, such as butter and meat, automobiles and electronics. The devaluation-cum-increased protection led to a reduction in imports and a rise in exports, resulting in an improvement in the trade balance, which moved from a deficit of $362 million in May 1991 to a surplus of $160 million in late September. But the initial gain was soon eroded.'[9] The crawl did not keep up with inflation; the real exchange rate appreciated about 30 per cent per year.

The trade balance in Poland did not improve in 1992. In December 1992, a 6 per cent across-the-board border (import) tax was imposed. According to Poland, the tax was imposed in view of balance of payments difficulties and the fall in official reserves due to the adverse consequences of the 1992 drought. The GATT Balance of Payments (BOP) Committee accepted this measure under Article XII.[10] In accordance with GATT rules, the decision was based on macroeconomic analysis and a statement made by the Fund Representative. It seems that fiscal considerations were equally important.[11] In 1992, the fiscal deficit widened to 7.0 per cent of GDP while the current account deficit reached 3.8 per cent and was not particularly high.

Poland initially expressed readiness to cut the tax to 3 per cent at the beginning of 1994, and to abandon the import tax by the end of 1994. In May 1994, Poland notified its intention of maintaining the tax to the end of 1996. During a subsequent round of GATT consultations in June of that month, the BOP Committee expressed its disappointment. As a result Poland committed itself to cut the import tax to 5 per cent at the beginning of 1995, then to 3 per cent in 1996 and to abolish the tax fully by the end of 1996. Indeed, the import tax was withdrawn on 1 January, 1997.

## Liberalization of trade resulting from European free trade agreements and the Uruguay Round

On 16 December, 1991, Poland signed the Europe Agreement (EA) (simultaneously with Czechoslovakia and Hungary) with the European Communities and their member states.[12] Following a two-year period, the EA took effect on 1 February, 1994. But the commercial part of the EA (the Interim Agreement: IA) had come into force on 1 March, 1992.

The EC and Poland started to create a free trade area (FTA) for non-agricultural products as far back as March 1994, to be completed over a maximum period of ten years. The timetable of liberalization is presented later in this chapter. The FTA is not applied to agricultural products, where liberalization is only limited and selective. In the case of European Union (EU) imports, this liberalization took five years and was completed by the end of 1997. For its part, Poland carried out a one-off tariff reduction involving 246 products by 10 percentage points following the coming into force of the IA.

The creation of the FTA is based on reciprocity and asymmetry for the EU; that is, requiring more rapid liberalization by the economically stronger partner (i.e. the EU).[13] Polish exports to the EU have benefited from a duty free treatment since 1 January, 1995, except for coal and steel, and textiles, duty free since 1996 and 1997, respectively.

The timetable of tariff liberalization of Polish imports was spread over time (see Table 5.1). Most reductions for other industrial products (43 per cent of Polish imports in 1993) were implemented in equal steps from 1995 until 1999.

The initial (1994) tariff rates were as follows:

- 5% – raw materials
- 10% – chemical products

*Table 5.1*  Poland: timetable of reduction of Polish tariffs in imports from the EU

| Date | Reduction of import tariff in the framework of EA (levels in percentage) | | | | | | | | |
|------|------|------|------|------|------|------|------|------|------|
| 1994 | 5 | 10 | 15 | 15 | 20 | 25 | 30 | 30 | 35 |
| 1 Jan 1995 | 3.4 | 6.9 | 10.4 | 12 | 13.8 | 18.9 | 22.6 | 23.0 | 26.4 |
| 1 Jan 1996 | 2.2 | 4.4 | 6.6 | 9 | 8.8 | 13.3 | 16.0 | 16.4 | 18.6 |
| 1 Jan 1997 | 1.2 | 2.4 | 3.6 | 6 | 4.8 | 8.3 | 10.0 | 10.4 | 11.6 |
| 1 Jan 1998 | 0.6 | 1.2 | 1.8 | 3 | 2.4 | 3.8 | 4.6 | 5.0 | 5.4 |
| 1 Jan 1999 | 0% | 0% | 0% | 0% | 0% | 0% | 0% | 0% | 0% |

*Source:  Raport w sprawie* (2000), p. 21.

- 15% – engineering industry products, paper and paper products, plastics, construction materials, steel products
- 15% – natural textiles, copper, some motor car accessories
- 20% – domestic appliances, footwear, products of glass and ceramics, some man-made fibres
- 25% – carpets and floor coverings
- 30% – hides, articles of apparel, other synthetic textiles
- 30% – final electronic products
- 35% – arms, ammunition.

The specific timetable applied to all other groups of products was as shown in Table 5.2.

From the inception of the IA, the standstill principle was to be applied to Poland's trade with the Communities. It means that no new import or export duties or other similar charges could be introduced, or the already existing tariffs and levies increased. The same refers to the quota restrictions and similar measures. 'The above regulations do not restrict agricultural policies pursued by Poland and the Communities, or any other measures applied within the framework of these policies.'[14]

Under the EA, the parties are allowed to restore some restrictions and implement new safeguards. Most of them are quite standard safeguard clauses, very similar to those that are a part of international trade regulations within the GATT system. They permit anti-dumping and countervailing measures (Art. 29), import restrictions in the case of balance of payments difficulties (Art. 64) or an unexpected rise in injurious imports (Art. 30), and export restrictions in the case of serious shortages (Art. 31).

On the other hand, there are some safeguard clauses reflecting specific features of the agreement and the principle of asymmetry. The most important is probably the so-called Restructuring Clause (Art. 28 of EA), which can be applied only by the Polish side in the form of increased import duties. 'These measures may only concern infant industries, or certain sectors undergoing restructuring or facing serious difficulties, particularly where these

*Table 5.2* Poland: timetable of liberalization of Polish industrial imports from the EU

| Commodity group | Share of Polish industrial products imports from the EU in 1992 (%) | Duration of liberalization |
|---|---|---|
| Appendix IVa – 1365 products, mostly investment equipment and raw materials | 25.0 | Liberalized completely in 1992 |
| Appendix IVb – cars and vehicles | 5.3 | 1992–2002. Since 1992, the duty free quota has increased annually by 5% for passenger cars and by 10% for trucks and cars equipped with catalytic converters. In 2002, the ban on cars older than 10 years and cars powered by two-stroke engines will be lifted |
| Appendix VI – some oils and *gases*, coal and coke as well as well as petroleum oils | 2.0 | Import licences were lifted at the end of 1996 |
| ECSC steel products – 8 eight-digit CN items – other products | 2.1 | Tariffs lifted in 1992, 1995–99 |
| ECSC coal products – 7 eight-digit CN items – other products | 11.6 | Tariffs lifted in 1992, 1995–99 |
| Textiles and clothes – 43 eight-digit CN item* – other textiles and clothes products | 10.9 | Tariffs lifted in 1992, 1995–99 |
| Other industrial products | 43.1 | Tariffs lifted 1995–99 |

*Source*: *Poland's Foreign Trade Policy 1993–1994*, 1994, p. 141.

*Notes*: * Taking into account the decisions of the European Council summit in Copenhagen on June 1993.

difficulties produce important social problems.' There are some important limitations in applying this clause. The customs duties 'may not exceed 25% *ad valorem* and shall maintain an element of preference for products originating in the Community'. These measures shall be applied for a period not exceeding five years and 'the total value of imports of the products which are subject to these measures may not exceed 15% of total imports of industrial products' from the EC. 'No such measures can be introduced in respect of a product if more than 3 years have elapsed since the elimination of all duties.'

Another specific provision is included in the article on competition and state aid (Art. 63 of EA). It states that any public aid which distorts or threatens to distort competition by favouring certain undertakings is incompatible with the proper functioning of the Agreement. But the '[p]arties recognize that during the period of five years after the entry into force of the Agreement, any public aid granted by Poland shall be assessed taking into account the fact that Poland shall be regarded as an area identical to those areas of the Community described in Article 92(3) of the Treaty establishing EEC'; that is, as a region in which the standard of living is low or the unemployment level high.

Finally, there are some special provisions in the so-called agricultural safeguard clause (Art. 21 of EA). The text reads that 'if, given the particular sensitivity of agricultural markets, imports of products originating in one Party, which are the subject of concessions. . . , cause serious disturbances to the market in the other Party, both Parties shall enter into consultations immediately to find an appropriate solution. Pending such solution, the party concerned may take the measures it deems necessary.' A very similar clause (Art. 14) is contained in the CEFTA agreement.

\* \* \*

Poland and EFTA member states (Austria, Finland, Ireland, Liechtenstein, Norway, Switzerland, and Sweden) concluded a free trade agreement in December 1992; that is, one year after signing the EA. It mainly covers trade in industrial products, but also encompasses some marine and processed agricultural products. EFTA members eliminated most of the obstacles to Polish imports upon the implementation of the agreement (in November of 1993). Under Poland's transitional period, tariff duties and quantitative restrictions on EFTA imports were phased out by 1 January, 1999 (except for steel, petroleum products, and automobiles). In addition, Poland has bilateral arrangements with individual EFTA members that either remove or lower tariffs on certain agricultural products.

Poland, along with the Czech Republic, Hungary and the Slovak Republic, established the Central European Free Trade Area (CEFTA) in 1992. Slovenia joined on 1 January, 1996, Romania on 1 July, 1997, and Bulgaria on 1 January, 1999. The primary objective of CEFTA was to establish a free-trade area by 2001, based on a system of bilateral liberalization schedules between members that adopted a framework of common rules. CEFTA covers all goods, except for a few agricultural products. Some 95 per cent of Poland's total tariff lines are covered by concessions under CEFTA. By 1996, almost 80 per cent of CEFTA trade in industrial products was tariff-free. By 1 January, 1999, tariffs had been abolished on almost all industrial products, except for certain cars, which were due to be removed by 1 January, 2002. Import quotas on industrial goods were also being phased out. Poland undertook efforts to abolish quantitative restrictions on all industrial products by

1997, except for automobiles and related products, where quotas would be removed by 2002. Although CEFTA does not specifically provide for free trade in agricultural products, parties have negotiated tariff concessions, subject to quotas, on such goods. Two tariff reduction schemes have been applied to agricultural products. In 1997, tariff preferences applied to four fifths of trade in agricultural products between the Czech Republic, Hungary, Poland, Romania and the Slovak Republic. Both free trade agreements (i.e. EFTA and CEFTA) provide the timetable of liberalization of Polish import tariffs for non-agricultural products fully compatible with the EA. The safeguard clauses included in all three agreements are also almost identical.[15]

In 1992, Poland signed bilateral free-trade agreements with the Baltic states (Estonia, Latvia and Lithuania), Israel and Turkey. These agreements provide for phased reductions of tariff duties on industrial goods, coupled with tariff rate quotas on a number of agricultural products.

In addition, Poland extends a Generalized System of Preferences (GSP) to 45 developing countries with a lower per capita GDP than itself, and to the 49 least developed countries (LDCs). Except for certain sensitive items, imports from developing countries are dutiable at tariff rates equivalent to 70 per cent of the MFN level. Imports from LDCs are duty free. Sensitive products excluded from these concessional schemes are certain agricultural, textile, and electronic products; tobacco; cosmetics; cars; and precious metals.

\* \* \*

Poland took part in the GATT Uruguay Round (UR) as the only state having the formal status of a developed country without any 'bound' customs duties. After submitting its initial offer on tariff concessions, Poland had bilateral negotiations with several countries. These concessions significantly reduced Polish import duties for non-agricultural products and were accepted by other members of the future WTO in December 1993. Poland's main commitments on trade in goods in the Uruguay Round were: (i) to bind, for the first time, close to 94 per cent of its tariffs; (ii) reduce tariffs by 38 per cent on industrial products and by 36 per cent on agricultural goods over six years; and (iii) to limit domestic support to agriculture by 20 per cent in value terms by the year 2000, and cut agricultural export subsidies by 36 per cent in value and 21 per cent in volume.[16] Thus, Poland became a founding member of the WTO on 1 July, 1995 on standard terms. Its special terms of accession to GATT became, therefore, irrelevant.

What was the impact of UR negotiations on Polish MFN import tariffs? The comparison of average, unweighted, and bound Polish and EU tariffs at the 2-digit HS level of aggregation after the UR is displayed in Figure 5.1.

In each HS group, the Polish non-agricultural import duties are significantly higher than those of the EU. The differences in each group decreased by 1999, after the implementation of the results of the UR negotiations.

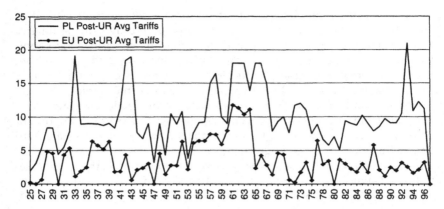

*Figure 5.1*   Poland: Polish and EU simple average bound tariffs after the UR
Source:   Maliszewska *et al.* (1998).

The simple average bound MFN Polish tariff rate for non-agricultural products was reduced from 16.73 per cent to 9.89 per cent.

This tariff disparity results partly from the strategy which Poland adopted during the UR. In general, it was assumed that the reductions offered to other WTO members should not diminish the level of final Polish bound tariffs beyond the level of the corresponding EU tariff after the implementation of the UR results. In this way, there will be practically no need for the EU to enter into compensation negotiations (based on Article XXIV:6 of GATT) as a result of the Polish accession.

The same general approach was, as much as possible, applied to agricultural products as well. Yet the situation was far more complicated because the new bound rates resulted from the 'tariffication', and the current level of EU tariffs is quite high. In the early 1990s, Poland increased MFN applied duties on a range of agricultural products and introduced variable levies. Consequently, with tariffication, Poland adopted relatively high 'bound' tariffs on many such products. Because Poland was a non-market economy for the base years of 1986–88 (selected under the Uruguay Round for estimating tariff equivalents of prohibited non-tariff barriers on agricultural products), it applied the higher EU tariff rates as the basis for tariffication.

*   *   *

What is the present, post-UR level of tariff protection in Poland? Recent data was provided during the WTO trade policy review of Poland in 2000 (see Table 5.3).

During the Uruguay Round, Poland increased its coverage of bindings from zero to 93.5 per cent of tariff lines. In agriculture, all tariffs were bound, compared with 92.2 per cent of lines for manufacturing. Poland adopted ceiling bindings for many of its products, especially in agriculture,

*Table 5.3* Poland: tariff indicators, 1999 (per cent)

| Indicator | All goods | Agriculture[1] | Manufacturing |
|---|---|---|---|
| Bound tariff lines[2] | 93.5 | 100.0 | 92.2 |
| Duty free tariff lines[3] | 2.5 | 0.2 | 3.2 |
| Simple average applied MFN rate[3] | 15.9 | 32.8 | 10.9 |
| Domestic applied MFN tariff 'spikes' (% of tariff lines over three times the overall tariff average, i.e. above 47.7%)[2] | 2.4 | 11.6 | – |
| Simple average applied preferential tariff (excluding developing country preferences)[4] | 6.0 | 9.2 | 5.9 |
| *Ad valorem* duties (per cent of tariff lines)[3] | 91.2 | 65.7 | 98.9 |

*Source:* *Trade Policy Review Poland*, Report by the Secretariat, 2000, Table III.

*Notes:* 1 In the case of agricultural products, there are also tariff quotas whereby much higher duties were set on above-quota imports. The Secretariat's estimates of tariff averages contained in this report used the above-quota rate. While this is likely to overestimate the rate of tariff protection afforded such products in cases where there are no above-quota imports, the Secretariat's tariff level estimates are also under-stated by using only the *ad valorem* component of mixed or composite duties. This was necessary because *ad valorem* equivalents were not available. For the same reason, a few specific duties, affecting only 30 tariff lines, were also excluded from the Secretariat's calculations.

2 Sectoral averages based on the WTO classification, which coincides with the definition of agriculture adopted in the Agreement on Agriculture.

3 Sectoral averages based on the HS definition.

4 Sectoral averages based on the ISIC definition. On an ISIC basis, Poland's average MFN tariff rate was 17 per cent on agriculture (and fisheries) and 16.1 per cent on manufacturing in 1999.

whereby bound rates considerably exceeded applied MFN rates on many products. Overall, the simple average bound tariff rate was 19.9 per cent, well above the average applied MFN rate of 15.9 per cent in 1999. This difference was greatest on agricultural products, where the bound average tariff rate was 55.5 per cent, almost two thirds higher than the average applied MFN rate (32.8 per cent) in 1999. This large difference was recently used in the late 1990s to raise applied tariffs to bound levels on a range of agricultural products (see the next section).

The economic importance of tariff protection is much better reflected by trade weighted tariff averages. This level of protection in Poland is substantially lower in comparison with unweighted (simple) tariff averages. It reflects the fact that a large proportion of Polish imports (over 70 per cent) fall under FTA agreements and GSP schemes. The data reflecting the scope of this phenomenon are presented in Table 5.4.

Data provided in Table 5.4 reveals the economic importance of the margin of preferences resulting from European FTA. These preferences are strongly pronounced in non-agricultural products, where almost all import tariffs (with the exception of transport equipment) are close to zero. On the other hand, the margin of preferences granted to the EU in agricultural products, in comparison with other countries, is not important.[17] These preferences are more important in imports from CEFTA countries. This data also reveals that the economic significance of tariff suspensions, which are quite often used by Polish authorities, is generally quite limited, although it can be important in some sectors.

In concluding, one should recognize that Polish import tariffs were substantially liberalized in the course of the 1990s. The level of tariff protection (measured by trade weighted averages) was 5.5 per cent in the second half of 1990 (when temporary suspensions were introduced) and 14.3 per cent in 1991.[18] In 1999, it was equal to 3.3 per cent, with tariff suspensions taken into account. Of course, this liberalization trend mainly reflects Poland's participation in free trade areas and, to some extent, in the UR. The liberalization is obvious in the case of industrial products. The trend in the level of agricultural protection was probably reversed.

\* \* \*

The Europe Agreement had the largest impact on Polish trade policy in the 1990s. What was the impact of the EA on Poland's trade flows and trade balance position? There have already been several studies made, analysing the scope of trade changes.[19] I cite here a recent study by W. Orlowski, analysing the impact of different stages of economic integration.[20] The summary of his estimates is presented in Table 5.5.

Orlowski, using a partial equilibrium model, with some elements of CGE, analysed both classic static and dynamic integration effects. He assumed that all dynamic effects (economies of scale, changes in the market structure, increased competition) are included in income effects, leading to an increase in the growth rate of per capita GDP. He also analysed 'adaptation' effects, leading to increases in production costs due to higher environmental protection standards, new technical standards, increased energy prices, and so on.

According to this study, the creation of the FTA with the EU boosted Polish exports by $2.4 billion and imports by $3.4 billion, and it reduced imports from third countries by $0.4 billion. For Poland, the consequences of adopting the common external tariff will be far more limited (increase of imports by $440 million – i.e. $163 + 389$). 

The dynamic effects appear to be much larger. The author assumed that after enlargement to the East, the EU would increase its GDP growth rate by

Table 5.4  Poland: average weighted import tariffs in Poland in 1999 (in %, according to the 1998 commodity pattern of imports)

| HS | Description | Without suspensions | | | | Including suspensions | | | |
|---|---|---|---|---|---|---|---|---|---|
| | | EU | CEFTA | Other countries | Total | EU | CEFTA | Other countries | Total |
| I-XXI | Total | 2.3 | 1.8 | 7.1 | 3.5 | 2.3 | 1.8 | 6.5 | 3.3 |
| I-IV | Agricultural products | 18.2 | 11.6 | 13.0 | 16.0 | 18.1 | 11.6 | 12.7 | 15.8 |
| V-XXI | On-agricultural products | 0.8 | 0.5 | 6.1 | 2.3 | 0.8 | 0.5 | 5.5 | 2.2 |
| I | Live animals and animal products | 26.2 | 15.1 | 13.9 | 22.8 | 26.2 | 15.1 | 13.9 | 22.8 |
| II | Vegetable products | 11.5 | 11.0 | 4.3 | 7.9 | 11.4 | 11.0 | 3.9 | 7.7 |
| III | Fats and oils | 18.5 | 3.9 | 12.6 | 13.7 | 17.6 | 3.9 | 12.4 | 12.8 |
| IV | Processed foodstuffs, beverages and tobacco | 20.4 | 15.9 | 23.0 | 20.5 | 20.4 | 15.9 | 22.7 | 20.4 |
| V | Mineral products | 0.7 | 1.7 | 0.9 | 1.2 | 0.7 | 1.6 | 0.8 | 1.2 |
| VI | Chemical products | 0.1 | 0.0 | 5.1 | 1.0 | 0.1 | 0.0 | 4.6 | 0.9 |
| VII | Plastics, rubber and related products | 0.00 | 0.0 | 7.5 | 1.3 | 0.0 | 0.0 | 6.8 | 1.1 |
| VIII | Leather, skin and related products | 0.00 | 0.0 | 16.6 | 3.5 | 0.0 | 0.0 | 15.9 | 3.4 |
| IX | Timber and timber products | 0.00 | 0.0 | 4.8 | 1.4 | 0.0 | 0.0 | 4.5 | 1.3 |
| X | Wood pulp, paper and board | 0.00 | 0.0 | 3.7 | 0.3 | 0.0 | 0.0 | 2.7 | 0.2 |
| XI | Fabrics and fabric products | 0.03 | 0.2 | 10.7 | 2.8 | 0.0 | 0.2 | 10.5 | 2.7 |
| XII | Shoes, hats, and cups | 0.0 | 0.0 | 14.9 | 6.7 | 0.0 | 0.0 | 14.9 | 6.8 |
| XIII | Glass, stone products, cement | 0.0 | 0.0 | 21.2 | 14.0 | 0.0 | 0.0 | 21.2 | 14.0 |
| XIV | Pearls, precious stones | 0.0 | 0.0 | 6.4 | 0.6 | 0.0 | 0.0 | 6.2 | 0.5 |
| XV | Non-precious metals | 1.3 | 1.6 | 8.9 | 2.4 | 1.1 | 1.6 | 7.4 | 2.2 |
| XVI | Machinery and equipment | 0.0 | 0.0 | 6.0 | 1.5 | 0.0 | 0.0 | 4.7 | 1.2 |
| XVII | Transport equipment | 6.1 | 1.3 | 12.0 | 7.7 | 6.1 | 1.3 | 11.7 | 7.6 |
| XVIII | Measure and controlling equipment | 0.0 | 0.0 | 6.7 | 2.9 | 0.0 | 0.0 | 6.4 | 2.8 |
| XIX | Arms and ammunition | 0.0 | 0.0 | 24.4 | 12.7 | 0.0 | 0.0 | 24.4 | 12.8 |
| XX | Various manufactures | 0.0 | 0.0 | 9.4 | 2.7 | 0.0 | 0.0 | 9.4 | 2.7 |
| XXI | Works of art | 0.0 | 0.0 | 0.0 | 0.0 | 0.0 | 0.0 | 0.0 | 0.0 |

Source:  Trade Policy Review Poland, Report by the Government, 2000, p. 10.

Table 5.5 Poland: static and dynamic effects resulting from Poland's integration with the EU. Changes in the trade balance position: cumulative effect of different stages of integration ($ million of 1996)

**Static effects**

| | Free trade area | | | Common external tariff | | | Common trade policy | | |
|---|---|---|---|---|---|---|---|---|---|
| | UE | Third countries | Total | UE | Third countries | Total | UE | Third countries | Total |
| Export to | 2348 | 0 | 2348 | 2348 | 0 | 2348 | 2348 | −34 | 2314 |
| Import from | 3398 | −389 | 3009 | 3398 | 163 | 3561 | 3398 | 214 | 3612 |
| Balance | −1050 | 389 | −661 | −1050 | −163 | −1213 | −1050 | −248 | −1298 |

**Dynamic effects**

| | Income effect in the EU | | | Income effect in Poland | | | Adaptation effects | | |
|---|---|---|---|---|---|---|---|---|---|
| | UE | Third countries | Total | UE | Third countries | Total | UE | Third countries | Total |
| Export to | 4442 | −34 | 4408 | 4442 | −34 | 4408 | 3344 | −383 | 2961 |
| Import from | 3398 | 214 | 3612 | 5047 | 858 | 5905 | 5620 | 1502 | 7122 |
| Balance | 1044 | −248 | 796 | −605 | −892 | −1497 | −2276 | −1885 | −4161 |

Source: Orlowski (2000), s. 66–71.

0.15 per cent and its imports by 0.7 per cent annually, for a period of five years. In Poland, the income effect should increase its GDP by 0.3 per cent, which should push up its imports by $2.3 billion. In total, there would be a large increase in trade volume without significant changes in the trade balance position.

The largest negative consequences might result from the adaptation effect. According to this study, it would reduce Polish exports by $1.4 billion and push up imports by $1.2 billion, thus substantially widening the country's trade deficit.

The estimates by Orlowski are based on some rigid assumptions and should be treated with caution. But it should be noted that almost all studies predict that integration with the EU will adversely affect Poland's trade position. This idea is also present in the Europe Agreement itself, in which the asymmetry of tariff liberalization and special safeguard clauses was introduced in order to diminish the economic burden of trade integration.

## Backsliding and the stability of Polish trade policy

The case study at the end of this chapter provides detailed examples of 'backsliding' in Polish trade policy in the 1990s. Here, I would like to present only the most significant examples of reversals in trade liberalization and discuss the possible causes of these developments.

The major trade policy developments, prior to signing the EA, EFTA, CEFTA agreements, and the Tokyo Round Final Act, have already been described. Only one additional event should be mentioned here.

In January 1992, shortly after signing the Europe Agreement, Poland raised customs duties for motor vehicles from 15 per cent to 35 per cent. At the same time, the duty free quota (for 30 000 vehicles) for automobile imports from the EU was granted under the Interim Agreement. India challenged these measures in GATT in November 1994 under Articles I (MFN clause) and XXIV (formation of customs union). Poland argued that these measures were introduced in order to curb excessive imports and that Polish tariffs were not bound in 1992. In November 1994, at a GATT session, a panel was set up on India's initiative to examine the consistency of the Polish system of automobile imports with the General Agreement provisions. According to India, the drop in sales of Maruti cars (small Indian-made automobiles) to Poland from 4 500 vehicles in 1991–92 to 500 in 1993–94 was due to: (i) the raising of customs duties at the beginning of 1992; and (ii) a considerable duty-free quota for imports from the European Union.

As it was impossible to conclude the panel proceedings by the end of 1995 (i.e. by the time of GATT expiration), in 1995 the Indian government withdrew its complaint from GATT 1947, and at the same time applied to Poland (September 1995) for consultations under Article XXIII: 1 GATT 1994 (i.e. under WTO). India argued that the Polish decisions violated the provisions of Article XXIV of GATT, including the text of Interpretation of Article XXIV of GATT 1994. In the opinion of the Indian government, the measures applied affected the advantages of this country agreed within GATT 1994. As a result, a number of bilateral meetings took place in order to find a solution satisfying both parties.[21] In the end, Poland agreed to open a temporary (two-year) tariff quota for small passenger cars originating in developing countries (within the system of special preferences – GSP). Such a solution allowed for a formal settlement of the trade dispute on the WTO forum.

* * *

The most notable examples of 'backsliding' trade policy in the second half of the 1990s involve the application of different safeguard clauses to Poland's imports from EFTA members.

On 25 August, 1994, the restructuring clause (Art. 30 of EA) was applied for the first time. Under the Europe Agreement, the customs duty on imported telecommunications equipment was abolished on 1 March, 1992, while the customs duty on components for this equipment remained effective. Its gradual liberalization was to be in effect from the beginning of 1995. A similar pattern of tariff concessions entered into force in connection with the free trade agreement with the EFTA and CEFTA countries. According to Poland, the decision to restructure and privatize the telecommunications industry required that the production of telecommunications equipment in Poland, based on component imports, should return to profitability.

On 21 September, 1994 customs duties on telecommunications equipment imported from the European Union, EFTA, CEFTA and Finland were also restored.

The imposition of this duty on telecommunications equipment met the formal requirements of the restructuring clause, since: (i) it referred to goods manufactured by a sector under restructuring; (ii) the new preferential duties did not exceed 25 per cent *ad valorem*; (iii) the preference margin has been maintained (12 per cent as compared to 15 per cent in imports from third countries); (iv) the overall value of imports subject to this safeguard measure did not exceed 15 per cent of total imports; (v) less than three years had passed since the abolishment of all the customs duties and quantitative restrictions; (vi) and a programme for the gradual phasing out of the revised customs duties had been adopted.

Since 1 January, 1995 the scope of the restructuring clause on telecommunications equipment has been extended. Namely, the level of duties on apparatus for carrier current cable systems originating in the European Union and the EFTA countries has been raised from 0 per cent to 8 per cent. At the same time, the list of telecommunications equipment tariff items – electrical equipment for telephony and telegraphy, teleprinters, telephone apparatus – was extended to include a fifth item, telegraphic apparatus.

The restructuring clause was also used in January 1996. Under these provisions, oil-refining products were exempt from the agreed timetable of liberalization. Poland extended the period of reducing customs duties on oil-refining products until 2001 (according to the original timetable, customs duties were to be cut to 0 per cent at the beginning of 1999). On 1 January, 1996, in connection with the restructuring clause, the duty applicable to petrol imports stood at 15 per cent (up from 12 per cent in 1995) instead of the 9 per cent envisaged by the agreement, while that applicable to diesel oil was 26 per cent instead of 21 per cent. The customs duty on fuel oil was increased to 25 per cent. In 1997, the customs duties on light and medium oils and gas oils amounted to 13 per cent and 20 per cent respectively.[22]

Poland applied the restructuring clause, for the third time, in January 1997. After consultations, Poland decided to keep tariffs at 9 per cent in order to protect its steel industry restructuring, in spite of the timetable liberalization timetable under the Europe Agreement. In 1998, the tariff rate was reduced to 6 per cent (against 3 per cent envisioned in the Europe Agreement), and in 1999 to 3 per cent instead of the planned 0 per cent. The clause was also applied to steel products imported from the Czech Republic and Slovakia, Hungary and Slovenia.

In all three cases, Poland met formal criteria included in the restructuring clause. In 1999, several steel and oil products were still covered by the restructuring clauses under the EA and CEFTA agreements. Duties on these products were not cut to zero on 1 January, 1999, and were higher than originally planned. The duties on steel products were cut to 3 per cent, from 6 per cent in 1998. In 2000, the duties were removed.

\* \* \*

Agriculture is another specific and very 'sensitive' sector with regard to which Poland extensively drew on special agricultural safeguards and other protective measures. Let it be recalled that, in July 1995, higher tariff rates were levied on imported foodstuffs and agricultural products in connection with Poland's membership of the WTO. The higher tariffs replaced all previous non-tariff measures applied to food and farm products. Some goods faced higher tariffs, including beef, some processed food products, yeast, sauces, alcohol and tobacco products.

On 28 January, 1997, the Polish government made a decision to restore customs duties on imported fodder cereals. The customs rate was 10 per cent (instead of 20 per cent prior to the suspension in spring 1996). Durum wheat, maize and soya were allowed in duty free. On 1 June, 1997, the import duty on barley was raised from 10 per cent to 20 per cent, and on 1 July, 1997, a 20 per cent import tariff on wheat came into effect. Also in 1997, under Article 14 the CEFTA agreement (agricultural safeguard clause), Poland withdrew tariff preferences for starch and starch products in trade with all CEFTA countries.

In January 1998 (under Art. 14 of CEFTA), Poland withdrew preferences for imports of sugar from the Czech Republic and Slovakia, seed and fodder maize from Hungary, fodder maize from the Czech Republic and Slovakia, and tomato concentrate from all Visegrad countries. The government took action under pressure from Polish farmers following a sharp increase in imports of Hungarian corn at the end of 1997. In July 1998, additional customs charges were set on imports of cut flowers and wheat seeds, and in October they were set on pork. These charges remained in effect until the end of 1998.

On 31 March, 1999 higher customs duties were applied for yoghurt, pig-meat, milk and wheat and meslin. Preferences for imports of yoghurt with fat content of 3 per cent to 6 per cent were withdrawn – the preferential 9 per cent rate applicable since 1 January, 1999 was replaced with an autonomous 35 per cent rate. Preferences (15 per cent) were withdrawn from wheat and meslin imported from the Czech Republic, Slovakia and Hungary and the autonomous rate was raised to the conventional rate level (from 20 per cent on 1 January, 1999 to 70 per cent). Other rises in customs duties involved pig livers, ossein and bones, parings and similar waste from raw hides and skins. The customs duty on malt used in the brewing industry was raised from 10 per cent to 20 per cent. In all cases, tariff increases were explained by excessive imports causing injury to domestic producers.[23]

In autumn 1999, a new round of political discussions began concerning agricultural protection in Poland. The Ministry of Agriculture requested many tariff increases blaming liberalization for a growing trade deficit in agricultural products and falling farmer incomes. The government was quite reluctant because of the adverse effects it might have on the accession negotiations and on inflation. Finally, in January 2000, import duties on flour made of other than wheat cereals, malt, bruised grain and bran were raised. The number of items on which duties were raised was substantially reduced in comparison with the original proposals.[24] Even so, the decision caused heated discussions and tension at meetings of the Association Committee.[25] The data in Table 5.6 summarize all safeguard clauses used by Poland.

Some additional comments to Table 5.6 seem relevant. First, the application of a safeguard clause can be regarded as a manifestation of backsliding in trade liberalization policy. But on the other hand, a delay in the liberalization timetable is a policy step explicitly foreseen by the letter of free trade agreements and as such cannot be treated as illegal or unfair. The same should be said about the European Union intensively using anti-dumping procedures against Polish imports.[26]

Second, in an overwhelming majority of cases, Poland met the criteria required for the application of a given safeguard clause. Probably the most ambiguous cases concern the agricultural sector. Here, the Polish government often acted under powerful pressure from farmers and lobbying groups, and criticism from the European Union was the strongest. Agriculture is very

*Table 5.6* Poland: safeguard clauses applied by Poland in trade with the EU, CEFTA, EFTA* and WTO in 1992–99

| Type of safeguard clause | Date of application by Poland |
| --- | --- |
| Restructuring clause (Art. 28 of Europe Agreement) (Art. 21 of EFTA Agreement) (Art. 28 of CEFTA Agreement) | 1 Customs duties on telecommunication equipment since 25 August, 1994. The list of telecommunications equipment for telephony and telegraphy teleprinters, telephone apparatus, was extended by a fifth item: telegraphic apparatus on 1 January, 1995. Higher duties were applied to imports from EU, EFTA and CEFTA countries. <br> 2 Customs duties and licensing of imports of petrochemical products since January 1996, *erga omnes*. <br> 3 In January 1997, Poland decided to maintain tariffs at 9% in order to protect its restructuring of the steel industry. The safeguard clause was applied with respect to imports from EU, Czech Republic, Slovakia, Slovenia, Hungary and Slovenia. |
| Anti-dumping clause (Art. 29 of the Europe Agreement) | Investigation started on 1 March, 1999 against a German exporter of X-ray film |
| General safeguard clause (Art. 30 of Europe Agreement) (Art. 20 of EFTA Agreement) (Art. 27 of CEFTA Agreement) | 1 From 21 July, 1994, *erga omnes* prohibition of imports of motor vehicles for transport of goods, bodies, chassis, trailers, coaches and special vehicles three years or older and used combines (since 1996 six years or older). <br> 2 From January 1996, imports of harvester combines were reduced to combines of six years or older. On 1 January, 1997 the prohibition was lifted. <br> 3 Between 24 August, 1996–1 July, 1997, temporary restriction (obligation to apply for a permit) of import of parts for industrial assembly of motor vehicles (internal combustion engines, chassis and bodies, accessories) <br> 4 On 31 March, 1999, higher customs duties were restored for yoghurt, porkmeat, milk and wheat and meslin. Preferences were withdrawn for wheat and meslin imported from the Czech Republic, Slovakia and Hungary and the autonomous rate was raised to the conventional rate level. Other rises in customs duties involved pig livers, ossein and bones, sinews, parings and similar waste from raw hides and skins. |

*Table 5.6* Continued

| Type of safeguard clause | Date of application by Poland |
|---|---|
| Safeguard clause in agriculture (Art. 21 of the Europe Agreement) bilateral with EFTA (Art. 14 of CEFTA Agreement) | 1 In 1997, Poland withdrew preferences for starch and starch products in trade with all CEFTA countries<br>2 In January 1998, Poland withdrew preferences for imports of sugar from the Czech Republic and from Slovakia, seed and fodder maize from Hungary, fodder maize from the Czech Republic and Slovakia, and tomato concentrate from Visegrad countries. In July 1998, additional customs charges were imposed on imports of cut flowers and wheat seeds, and in October they were set on pork. These charges remained in force at the end of 1998. |
| Re-export and serious shortages clause (Art. 31 of the Europe Agreement) (Art. 23 of EFTA Agreement) (Art. 29 of CEFTA Agreement) | 1 Prohibition of exports of various skins – between 4 September, 1993 and 31 December, 1993; from 1 January, 1994 to the end of 1995 – prohibition of exports of skins and hides wastes, and an export quota for skins and leather was established.<br>2 From 1992, temporary restrictions on exports of metal, steel and non-ferrous metals wastes. |
| 'General exceptions' clause (Art. 35 of the Europe Agreement) | Retaliatory measures on imports of even-toed animals and products thereof from the EC (as those applied by the Community between 7 April, 1993 and 13 July, 1993) |
| Balance of payments difficulties clause (Art. 64 of Europe Agreement) (Art. 24 of EFTA and 32 of CEFTA) | 17 December, 1992 – *erga omnes* import tax (6%), lowered to 5% on 1 January, 1995, to 3% in January 1996 and abolished on 1 January, 1997. |

*Source*:  E. Kaliszuk and E. Synowiec (1995) 'Klauzule ochronne z Ukladu Europejskicgo: gdzie zastosowano i w czym pomogly obu stronom', *Rynki Zagraniczne*, no. 24, February 25, and annual reports (*Foreign Economic Policy of Poland*) published by the Foreign Trade Research Institute in Warsaw.

*Notes*:  \*Safeguard measures provided for by this clause were also applied to imports from the EFTA and CEFTA countries under the relevant free trade agreements concluded with these countries.

sensitive in the EU and in many other European countries.[27] But even in this sector, Poland never raised its applied import tariff above the level of bound duties contained in the WTO Schedule of concessions.

Third, it seems that in most cases (perhaps with the exception of agriculture) the extent of safeguard clauses was quite limited in time and scope.

It probably had no significant impact on bilateral trade flows, especially if we compare the scope of the clauses with Orlowski's estimates of trade imbalances induced by the Europe Agreement. Yet, it does not mean that it had no impact on Polish–EU trade relations and negotiations.

## Determinants of Polish trade policy

In assessing trends in trade policy, we should limit the analysis to those sectors in which safeguard provisions were applied.

The most protection prone sector is obviously agriculture. This sector is very sensitive almost everywhere in Europe. It is a truism to say that agriculture has always attracted attention and provoked tensions in west European integration. The reform of CAP has been, and still is, one of the most difficult tasks.

In the case of Poland, this sensitivity is easy to understand. We should recall that in the mid-1990s, 27 per cent of the Polish population was still employed in the agricultural sector, and the level of unemployment in rural areas was well above the Polish average (13 per cent). The agricultural lobby in Poland remains quite well organized; Polish farmers have an important parliamentary representation and are able to organize major protest actions (blockades of roads, demonstrations in front of the Parliament, and so on). The protectionist lobbyists argue that Poland, despite its comparative advantage in agricultural production, runs a trade deficit in these products. They claim that this situation is primarily a result of major export subsidies provided to EU producers, and they call for financial support and protection of domestic farmers.

The legal situation enabled changes in trade policy. The level of Polish 'bound' duties in the WTO is quite high and fully comparable to that of the EU, while the applied agricultural duties are substantially lower. Hence, the government had leeway with respect to agricultural duties. The potential limitation resulted rather from provisions of the EA and other FTA agreements.

There were obvious conflicting policy goals in this area. The government was tempted to use low duties (or, at least, not to increase them) in order to reduce inflationary pressures in Poland. On the other hand, higher duties were perceived as necessary to appease the protectionist lobby, and they could be desirable in order to reach a level of external protection comparable to that in the EU. Gradually, reaching the same level of external protection will help to avoid any possible compensation negotiation resulting from accession to the EU (GATT Art. XXIV).

The government, acting under strong pressure, occasionally yielded to the protectionist agricultural lobby. But each increase in customs duties, being applied to imports originating from the EU as well, caused tension in relations with the European Union and could adversely affect the accession negotiations. The unclear terms and an unspecified timetable of accession seriously weakened the government's position in its dealings with agricultural lobbies.

*   *   *

The economic situation of the sectors in which restructuring clauses and other forms of protection were applied is diversified.

The motor vehicles sector initially received the highest level of protection. The tariff rates, which are unbound in the GATT/WTO, were raised in 1992, provoking a dispute with India. A special delayed timetable of import liberalization was foreseen in the EA and other European FTA agreements. As compensation, tariff free quotas for passenger cars and trucks imported to Poland were gradually introduced.

The government provided protection with a view to restructuring that the industry badly needed in order to improve competitiveness. The high level of protection on final goods was combined with duty free imports of CKD kits and favourable tax arrangements. Thus, foreign manufacturers developing assembly operations in Poland enjoyed a very high level of effective protection. Consequently, motor vehicle production expanded significantly and attracted about 23 per cent of foreign investment. In 1998, it accounted for 8 per cent of manufacturing output and about 7 per cent of Polish exports.[28] Daewoo and FIAT bought two major car factories in Warsaw and Bielsko-Biala, respectively, while some other foreign companies (notably, Opel) started their operations in Poland.

The economic efficiency of this policy is not easy to estimate. But the increases in efficiency and production volumes were quite encouraging. The government considered the motor car industry to be strategic, as in many EU countries.[29] The fast rising domestic demand, after a period of non-market shortages, provided a strong stimulus for the sector. Therefore, it is not easy to demonstrate that protectionist policy, based on the infant industry argument, was ineffective here.

The petroleum sector is also in need of substantial restructuring. This sector is very capital intensive and needs large investment in modernization.[30] The potential competition from large MNEs is also very strong in this sector.

Most crude oil is imported from either Russia or Kazakhstan and then refined locally. Approximately 20 per cent of Poland's refined product requirements is imported. Imports of these products are subject to MFN rates of 15 per cent, 20 per cent, or 25 per cent. However, some refined products, including aviation fuel, were subject to specific rates of either 25 per cent or, if higher, ECU 25 per ton, and 35 per cent or, if higher still, ECU 33 per ton.[31]

The genuine privatization started in 1999. The Polish Oil Concern PKN, which controls some 70 per cent of the oil market and 40–50 per cent of the retail market, was partially privatized in October 1999. Thirty per cent of its stock was privatized, with just over half sold to foreign investors. Splitting the state sector and separating the refinery process from storage and pipeline distribution of oil and gas has facilitated the industry's privatization.

Additional temporary protection has helped to raise the profits of Polish refineries and probably helped privatization. The restructuring process will continue for several years. On the other hand, the rise in world crude oil prices forced governments to suspend duties on refined products in September 2000 in order to diminish the cost-pushed inflationary pressure.

Another sector benefiting from protection (in fact, a temporary delay in liberalization) is the iron and steel industry, which is undergoing an important restructuring programme. Poland is the nineteenth largest world producer and one of the biggest employers (90 000 employees in 1987) in the region of Silesia. This industry accounts for some 13 per cent of domestic industrial output, and a quarter of its production is exported.

Under the restructuring programme, new privatization plans for all steel mills commenced in 2001. However, further privatization of the industry remains a major challenge, given the consistently high share of production from state-owned enterprises and their dependence on state aid. In this connection, the government also approved a labour redundancy scheme costing Zl 12 billion ($4 billion), aimed at reducing employment to 40 000 over the plan's life.[32] This programme is also important from both social and ecological points of view.

It appears that the European Union exercised strong pressure for the sector's restructuring and privatization prior to accession.[33] On the other hand, the EU quite often used anti-dumping procedures against Polish imports of steel products.[34] Poland insisted on delaying the liberalization of import tariffs. It seems that the application of the restructuring clause, providing for a one-year delay in liberalizing Polish imports, was a reasonable exemption, given the scope of challenges faced by the iron and steel sector in Poland.

The last sector benefiting from the restructuring clause was the telecommunications equipment industry. Although it has grown in recent years, it still requires additional investment and modernization to foster economic development. Poland has a relatively low density of fixed telephone lines. Network investment and penetration rates in Poland were among the lowest in the OECD in 1997, with price levels the second highest. The telecommunications sector is expected to grow at about 25 per cent per annum, and was expected to reach $5 billion in 2000.[35] Poland's telecommunications market was dominated by TPSA, the state-owned national operator. It had a statutory monopoly of international calls, telex and telegraphic services. Competition in local and domestic long-distance calls was allowed. But TPSA effectively held a monopoly of domestic telephone services, owning 99 per cent of local lines.

However, in recent years, the government has adopted a more vigorous policy aimed at the privatization and deregulation of the telecommunications sector. A new Telecommunications Act, intended to be adopted in 2000, was enacted to remove TPSA's monopoly of telex and telegraphic services. Its statutory monopoly on international calls and mobile satellite services and network ended in line with Poland's GATS commitments in 2003. The government commenced privatizing TPSA in late 1998, with a successful flotation on the Warsaw Stock Exchange and the sale of 15 per cent of the shares (5 per cent were offered to domestic investors and 10 per cent overseas). The government intended to privatize another 25 per cent to 35 per cent and distribute a further 15 per cent to employees by the end of 1999 and strategic investor has been chosen.

The international competition in this sector of services is particularly strong today. Until recently, telecommunications services in most west European countries were dominated by one domestic national operator. A temporary safeguard clause on telecommunications equipment was aimed at protecting Polish equipment producers in a strongly growing sector. Presumably, it was perceived as a complement to granting protection to the TPSA. The production of telecommunications equipment is technology intensive and reveals learning effects and economies of scale. Thus, the protection was probably regarded as an element of shielding a modern sector, which might demonstrate a positive external effect in the long run, although it has adversely affected the short-run cost of modernization of telecommunications services.

## Conclusion

A very clear trend towards the liberalization of trade policy and the foreign exchange system could be seen in Poland in the 1990's.[36] During the first stage of transformation Poland, as some other transition economies, relied on macroeconomic policy instruments.[37] Around 1990, the scope of tariff liberalization was initially moderated by an overvalued domestic currency, and from 1992 by the application of across-the-board import taxes. The second stage of genuine liberalization started with the implementation of free trade agreements with the EU, EFTA, and CEFTA countries and was reinforced by the conclusion of Uruguay Round negotiations. The final stage started with the accession to the European Union and adoption of the common external tariff. The present level of trade weighted protection (3.3 per cent in 1999) is low, reflecting the provisions of the EU, CEFTA, EFTA agreements and the country's WTO commitments. The Europe Agreement is particularly important because it covers approximately 67 per cent of Polish imports.

The pace of liberalization was well defined for non-agricultural products. Here, the upper limits for MFN tariffs are well defined in the WTO schedule and are reduced to zero in the framework of FTA agreements. An overwhelming majority of MFN tariffs are bound, with the notable exception of motor vehicles. In the case of free trade agreements, the obligations to remove tariffs can only be modified by recourse to restructuring or safeguard clauses.

These trade policy measures cause some tension with trade partners and provoke discussions on how to interpret the standstill provision.

The situation is less clear in the case of agricultural products, despite the fact that 100 per cent of Polish tariffs are bound. The scope of trade liberalization within free trade areas was – in most cases – limited while WTO bound tariffs in Poland were significantly higher than the applied ones. This situation gave leeway in Poland's trade policy and sometimes caused significant tensions in relations with its partners.

In accordance with orthodox trade theory, liberalization can and should be made autonomously. But the reduction of tariffs, if incorporated in the schedule of commitments, reduces the government's leeway and diminishes its bargaining power in future negotiations. It can also provoke requests for compensation in the accession process to the European Union (Art. XXIV of GATT). Poland, as other countries, was very rarely making permanent autonomous commitments.

Instead, tariff suspensions and tariff quotas were frequently used in Poland in order to limit shortages of goods, increase the pressure on domestic producers and, above all, to decrease protection and prices of raw materials and components for manufacturing industries. The latter was also treated as a necessary instrument to increase the level of effective rate of protection for some manufacturing industries (i.e. motor cars).

Tariff suspensions and tariff quotas are measures of trade liberalization, but they can introduce an element of discretion and instability in trade policy. Thus, they can potentially create rent-seeking activities and cause allegations of lack of predictability in trade policy. The accession to the EU is expected to reduce – at least, to some extent – the scope of this phenomenon.[38]

The pattern of sectoral protection and its economic 'rationality' is not analysed here. The purpose of this chapter is more limited, and the analysis confined to changes in trade policy. Special attention is devoted to examples of backsliding within the framework of trade preferential agreements.

What sort of arguments can explain the pattern of sectoral backsliding in Poland? At first glance, it seems, the government (perhaps rather intuitively) considered strategic trade policy arguments in choosing sectors for temporary protection. Presumably, in all protected manufacturing sectors described here, the international competition is imperfect, products are differentiated, and economies of scale are significant.[39] Therefore, arguments in favour of strategic trade policy could, in principle, be relevant. But their relevance is based on many rigorous assumptions and empirical support for these theories is rather weak.[40] More important is the fact that these arguments are almost irrelevant in the case of small producers that do not have a dominant position on the world market. Therefore, they do not represent valid arguments for protection in the case of Poland. Another simple argument states that extended tariff protection of a large market may attract foreign direct investment (FDI).[41] It seems that this argument could only have been important in the case of the motor car industry in Poland. In this case, the scope of protection was large and the inflow of FDI significant (23 per cent of the total).

Therefore, being economically rational and observing international obligations, the Polish government should not use trade policy to protect the above-mentioned manufacturing sectors in future – especially because privatization and restructuring is already well under way in all these sectors. Another important limitation results from the Europe Agreement itself. The restructuring clause most frequently used by Poland, cannot 'be introduced in respect to a product if more than three years have elapsed since the elimination of all duties . . . '. It means that very soon it will be impossible to introduce a new restructuring clause.

It must be stated that the evolution of trade policy on agricultural products is a special case. The political economy was the most important determinant of trade policy in this sector. This happens in many other European countries. Since the mid-1990s deterioration in a farmers' standard of living and rising unemployment in Poland have become evident. It was especially visible in the regions in which large state-owned farms have gone bankrupt. The political pressure from well organized protectionist lobbies has increased strongly since that time. The government, under strong pressure, sometimes conceded to these lobbies, especially because the farmers' parliamentary representation was always very strong.[42] The accommodating stance of the government was facilitated by relatively weak external legal constraints. The level of bound duties was quite high, well above the applied level. Therefore, it was possible to raise the level of tariffs without breaching WTO commitments. On the other hand, the provisions of a special agricultural clause (Art. 21 of EA) were not very precise. Their relationship with the standstill provisions is slightly ambiguous. And there are no special market preconditions for application of the clause. It seems that at the time of EA negotiations, both sides were interested in having vague language in this clause.

Each rise in customs duties, particularly those applied to imports originating from the EU, caused tension between the European Union and Poland and could adversely affect accession negotiations. The situation is again expected to improve with accession. What is most encouraging is an agreement concluded in September 2000 which widened the scope of liberalization in agricultural trade between the EU and Poland. According to press reports, 75 per cent of bilateral trade has been liberalized.[43] For non-sensitive products, such as fruits, vegetables or horsemeat, the trade is fully liberalized and no new restrictions can be imposed. For sensitive products – such as pig meat, poultry, butter, cheese flour or wheat – duty free quotas were established and should be increased by 10 per cent each year.[44] Poland reinstalled the previously applied tariffs (within traditional tariff quotas) for malt, sugar and rape seed. No new import restrictions can be imposed on these products and neither side can subsidize their exports. The latter was a frequently repeated demand by Polish farmers.

The September 2000 agreement has reduced tensions between the Commission and Poland, and facilitated the agricultural branch of accession negotiations. The demands for protection, coming from agricultural lobbies, should diminish. The situation should become more predictable after the accession. The unclear terms and an undecided timetable of accession weaken the position of the Polish government in its dealings with agricultural lobbies.

## Case study: the escape clause used by Poland in 1994 in the telecommunication equipment sector; simple estimates of economic implications

Poland applied the restructuring clause in the framework of the Europe Agreement (EA) for the first time in August 1994. The clause was used with regard to telecommunication equipment. In 1994, customs duties on imports from the European Union were eliminated, while import duties from rest of the world were close to 15 per cent.

Poland invoked Article 28 of EA and increased its duties on imports from the EU to 12 per cent. As a result, the changes shown in Table 5CS.1 occurred in mid-1994:

The Armington (1969) model was used to estimate possible trade, production and welfare changes resulting from the safeguard clause. In order to simplify the task, I adopted the standard version of the Armington model as elaborated by Francois and Hall (1997).

*Table 5CS.1* Changes in import tariffs resulting from the application of the safeguard clause to the telecommunication equipment sector in 1994

| CN commodity number | MFN duties on imports from developed countries – WTO member states in 1994, 1997 and 1999 | | | Duties on imports from EU according to EA in 1994 | Duties on imports from EU after invocation of Article 28 of EA in 1994 |
|---|---|---|---|---|---|
| 851710 | 15 | 8.4 | 8 | 8 | 8 |
| 851720 | 15 | 5 | 8 | 0 | 12 |
| 851730 | 15 | 8 | 8 | 0 | 12 |
| 851740 | 15 | 8 | 5 | 0 | 12 |
| 851781 | 15 | 8 | 8 | 0 | 12 |
| 851782 | 15 | 8.7 | 8 | 0 | 12 |
| 851790 | 15 | 8 | 7 | 8 | 8 |

Using their terminology we distinguish a composite good $q$ consisting of different products $(X_i)$, differentiated by their place of production. Following Armington, we assume well behaved preferences over a weakly separable product category that comprises similar but not identical products. These imperfect substitutes are differentiated by their country of origin. The demand for a composite good $q$ is given by

$$q = \left[ \sum_{i=1}^{n} \alpha_i X_i^{\rho} \right]^{1/\rho} \tag{1}$$

where $\alpha_i$ is a weight of product $i$ in expenditure for good $q$ in accordance with CES function and $\rho = 1 - \left( 1/\sigma \right)$, ($\sigma$ being the elasticity of substitution between products differentiated by the place of origin).

Under these assumptions the price index for the composite good is

$$P = \left[ \sum_{i=1}^{n} \alpha_i^{\sigma} P_i^{1-\sigma} \right]^{1-1/\rho} \tag{2}$$

and from first-order conditions we get demands for $X_i$ products

$$x_i = \left[ \frac{\alpha_i}{P_i} \right]^{\sigma} \left[ \sum_{i=1}^{n} \alpha_i^{\sigma} P_i^{1-\sigma} \right]^{-1} Y = \left[ \frac{\alpha_i}{P_i} \right]^{\sigma} P^{\sigma-1} Y; \tag{3}$$

where $Y$ is the income spent on composite good $q$ and $P_i$ is the price of product $X_i$.

Defining the linear supply equation as $K_{si} P_i^{\varepsilon_{si}}$, where $\varepsilon_{si}$ is a constant supply price elasticity and $K_{si}$ is a constant, we receive the excess demand conditions in each product $i$ as follows:

$$\left[ \frac{\alpha_i}{P_i} \right]^{\sigma} P^{\sigma-1} Y - K_{si} P_i^{\varepsilon_{si}} = 0 \tag{4}$$

Rewriting the price index equation (2) into form:

$$\left[ \sum_{i}^{n} = \alpha_i^{\sigma} P_i^{1-\sigma} \right]^{1-1/\rho} - P = 0 \tag{5}$$

and defining the demand for the composite good $q$ as

$$q = k_A P^{NA} \tag{6}$$

where $NA$ is the elasticity of demand for the composite good, we can specify the market clearing condition for the composite good as

$$k_A P^{NA} - Y = 0 \tag{7}$$

Taking the relevant elasticities from the other models, calibrating the wages ($\alpha_i$) and constant term of the composite demand function, equations (4), (5) and (7) define a system of $(n+2)$ equations and $(n+2)$ unknowns. The system can be solved for the prices of $n$ products ($P_i$), income ($Y$) and the price index of the composite good ($P$).

In order to analyse changes in import tariff for good $j(t_j)$, we modify equation (4) and get

$$\left[\frac{\alpha_i}{P_i}\right]^{\sigma} P^{\sigma-1} Y - K_{si}\left(\frac{P_j}{1+t_j}\right)^{\varepsilon_{sj}} = 0 \tag{8}$$

In our case, we assume that product $X_1$ is produced domestically while $X_i$ are imports from countries $i = 1, 2$. Changes in prices in products from preferential European countries (origin 2) and from the rest of the world (origin 3) are modified by import tariffs.

*   *   *

Several simplifying assumptions were necessary in order to estimate the implications of the safeguard clause. The CN 8715 commodity group, covered by the safeguard, contains 'electrical apparatus for line telephony or line telegraphy, including such apparatus for carrier-current line system'. Two tariff lines were excluded from the safeguard coverage, namely CN 851710 (telephone sets for line telephony) and CN 851790 (parts for electrical apparatus for line telephony or telegraphy).

The production data received from the Polish Central Statistical Office are at the 3-digit level of NACE Rev.1 classification. This level of NACE classification is more aggregated than 6-digit level of CN classification, used for trade purposes.

In the newest edition of EUROSTAT statistics (COMEXT), the CN 8517 group is included in the sector NACE 344 (telecommunication and measuring equipment). In the Polish version of NACE (PKD), the NACE 322 sector roughly corresponds with the CN 8517 commodity group.[45] But it was impossible to use the whole NACE sector, given the fact that products CN 851710 and 851790 were excluded from the safeguard provisions. Therefore, I assumed that the share of these 6-digit commodity lines in total production of the sector (NACE 322) is the same in imports and in production.[46] The production data used for simulation are presented in the Table 5CS.2.

The elasticity of substitution for this sector (2,11) is taken from Shiells et al. (1986) and other parameters are 'reasonably' assumed. The price elasticity of EU and ROW import supplies equals 1.5, while elasticity of domestic supply equals 3. The composite elasticity of demand is assumed to be 1.5. The results of the simulation are presented in Table 5CS.3.

The results presented here reflect two variants of the model. The first column reports the sectoral Armington model as described in equations

*Table 5CS.2*  Production and trade data used for simulation using Armington model (in $ thousand of 1994)

| Sector 322 NACE rev.1 | Domestic production | Domestic production for domestic market | Exports of domestic producers | Total imports | Imports from the European Union |
|---|---|---|---|---|---|
| | 211133 | 189845 | 21288 | 26068 | 22022 |

*Table 5CS.3*  Results of estimation according to Armington model

| Changes in major variables resulting from application of the safeguard clause | Non-linear sectoral Armington model | Sector focused Armington computable general equilibrium model |
|---|---|---|
| per cent change in internal price of domestic good | 0.13% | 0.3% |
| per cent change in shipments of domestic good | 0.39% | 0.8% |
| per cent change in border price of goods from EU | −1.35% | −0.1% |
| per cent change in internal price of goods from EU | 10.49% | 10.5% |
| per cent change in imports of goods from EU | −18.44% | −17.9% |
| per cent change in border price of goods from ROW | 0.04% | 0.0% |
| per cent change in internal price of goods from ROW | 0.04% | 0.1% |
| per cent change in imports of good from ROW | 0.58% | 0.0% |
| Changes in thousand $US | | |
| Net welfare effect | −1997 | 42* |
| Change in tariff revenue | 2130 | 2164 |
| Change in domestic production | 739 | 1501 |
| Change in imports form EU | −2309 | −3954 |

*Note*:  *Change in equivalent variation-based measure of change in welfare.

(1)–(8). The second column represents results of a variation of Armington model, calculated in general equilibrium framework (CGE). The rest of the economy, represented by the volume of GDP, is treated here as a *numeraire* sector. In this framework, changes in the telecommunication equipment production have explicit feedback implications for other sectors of the economy. Therefore, production changes are more pronounced.

The results of two variants of the same model are not very different. In both cases, the increase in tariffs raised on EU imports increases the price of goods imported from this region by 10.5 per cent and causes a significant decrease in the volume of imports. It also raises, to some extent, prices of goods produced domestically and imported from the rest of the world (ROW). The price impact is limited because goods are imperfect substitutes and supplies from both sources rise.

Domestic production in the first variant rises by 0.4 per cent ($739 thousand) and by 0.8 per cent ($1.5 million) in the CGE model. This was the official reasoning for introducing the safeguard. The introduction of the safeguard also raises tariff revenue by about 2.1 million US$. On the other hand, according to the partial equilibrium model, it decreases net national welfare by $2 million. In the case of the CGE variant, the net welfare is slightly increased due to feedback effects in other sectors of the economy. Still, we may conclude that production and welfare effects are not trivial but are fairly small.

Two additional comments are necessary. The results presented here form an upper estimate due to the fact that the product coverage of NACE 322 is larger than that of CN 8517. Conversely, the Armington model does not capture theoretically possible gains from economies of scale. According to other studies, the increasing returns to scale are quite important in the telecommunication sector.[47] They should increase the possible efficiency effects.

Simulations presented here are indicative only. A calibration process and the application of parameters from other models are contested by some economists. The model is sensitive to assumptions made about elasticity of substitution. But there are only very few other estimates of this parameter. On the other hand, it is quite insensitive to changes in demand and supply elasticities. For example, changing the price elasticity of domestic supply from 3 to 2 reduces the increase of domestic supply from $739 million to $630 million, leaving other results almost unaltered. Therefore, we can conclude that the results presented here should at least reflect a range of changes.

## Notes

1 Kierzkowski *et al.* (1993), pp. 30–1.
2 At that time, the Hungarian inflation was equal to 18.9 per cent while in Czechoslovakia it was around 1.5 per cent. See Mayhew (1998), p. 204.
3 Mayhew (1998), p. 246.
4 Kostecki (1979).
5 During the last week of December 1989, the official rate stood at 5,560 zl./$, the average action rate was 6,000 zl.$ and the parallel at 8,000 zl./$. Kierzkowski *et al.* (1993), p. 31.

 6  Kierzkowski *et al.* (1993), p. 31–2.
 7  Ibid. p. 48–9.
 8  *Poland's Foreign Trade Policy 1993–1994* (1994), p. 63.
 9  Kierzkowski *et al.* (1993), p. 56.
10  GATT document: BOP/R/206.
11  This approach is in accordance with theoretical arguments presented by Begg in this volume. He argues that import tariffs may improve tax capacity and competitiveness and thus support the macroeconomic stabilization programme of a country at an early stage of economic transition.
12  *Europe Agreement* (1994).
13  The principle of asymmetry, mirroring from the infant industry argument, was reflected also in some safeguard clauses.
14  *Poland's Foreign Trade Policy 1993–1994* (1994), p. 139.
15  A detailed comparative analysis of differences between safeguard clauses included in EA and EFTA and CEFTA agreements is provided by Synowiec (1995).
16  *Trade Policy Review Poland*, Report by the Secretariat, 2000, p. 24.
17  In fact, the average import tariff for agricultural products is higher in imports from the EU in comparison with other countries (18.2 per cent against 13.0 per cent). This probably reflects the fact that the level of protection is higher for temperate zone products (in which Poland and EU are producers) than for tropical zone products.
18  *Poland's Foreign Trade Policy 1993–1994* (1994), p. 63.
19  For example, *Umowy o wolnym handlu* (1995), Maliszewska *et al.* (1998) or Marczewski (1999).
20  Orlowski, 2000.
21  WTO Document, WT/DS/19/2, dated 11 September, 1996.
22  *Poland's Foreign Trade Policy, 1995–1996*, p. 58–9.
23  The detailed list of customs changes is provided in *Zagraniczna polityka gospodarcza* (2000), pp. 311–13.
24  Ibid. p. 314.
25  The controversy is whether a country can raise its import duty (invoking a safeguard clause but contrary to the standstill provision) on a product on which no concessions were granted under FTA agreement. The EU presented this sort of reading of the EA at an early stage of EA operation. In 1999–2000, Poland, contrary to the EU, advocated for this interpretation.
26  There were more than twelve anti-dumping (A-D) procedures started against Polish exporters in the EU in the 1990s. In 2000 nine anti-dumping duties were applied against Polish exporters (*Zagraniczna polityka gospodarcza i handel zagraniczny Polski 1999–2000* (2000), p. 328.
27  For example, some countries (including Argentina, Australia, New Zealand and the USA) claimed that Hungary violated the WTO Agreement on Agriculture by providing export subsidies in respect of agricultural products not specified in its schedule and by providing subsidies in excess of its commitment. On 30 July, 1997 parties to the dispute notified the WTO that they had reached a mutually agreed solution (WTO document WT/DS35).
28  *Trade Policy Review Poland*, Report by the Secretariat (2000), p. 104–5.
29  European countries used different forms of external protection mainly against Japanese producers. The most notable were Voluntary Export Restraints (VERs) imposed on car exporters to the UK, France and Italy.

30  The Government's plans for the sector were contained in the 1995 Law on Restructuring and Privatization of the Petroleum Sector.
31  Lower preferential rates apply to imports of refined products from the European FTA. For example, imports of diesel oil from the EU, EFTA and several CEFTA partners were dutiable in 1999 at 11 per cent, and 17 per cent if imported from Bulgaria or Romania (*Trade Policy Review Poland*, Report by the Secretariat (2000) p. 89).
32  *Trade Policy Review Poland*, Report by the Secretariat (2000) p. 103.
33  EBRD (1999).
34  In 1999–2000, five (out of nine) A-D duties imposed in the EU against Poland involved iron and steel products. See *Zagraniczna Polityka Gospodarcza i Handel Zagraniczny Polski 1999-2000* (20000), p. 328.
35  *Trade Policy Review Poland*, Report by the Secretariat (2000), p. 109.
36  According to EBRD Poland, the Czech Republic, Hungary, Bulgaria, Slovakia and Slovenia reached 4+ level of trade and foreign exchange system liberalization. The scale is from 1 (little progress) to 5 (standards and performance comparable to advanced industrial countries) (EBRD, 1999, p. 24).
37  Kawecka-Wyrzykowska (2000) p. 9–12.
38  In fact, the European Union also uses temporary suspensions and tariff-free quotas, to some extent.
39  Main arguments in favour of strategic trade policy (shifting profits to domestic producers) are discussed in Eaton and Grossman (1986), Krugman (1984).
40  See, for example, Krugman and Smith (1994).
41  This argument was frequently used in the 1960s when the protected EC market attracted large inflows of American FDI.
42  The Polish Peasants Party (PSL) participated in the governing coalition from 1994 to 1997.
43  Rzeczpospolita, 28 September, 2000. The agreement entered into force on 1 January, 2000. Similar agreements were concluded with other East European candidate countries.
44  The duty free quotas are not always symmetric. In some cases, like pork meat, the Polish export quota is slightly larger than that of the EU.
45  The 344 NACE sector does not exist in the NACE Rev.1 classification. The 322 NACE contains more products than the CN 8517 commodity group. Therefore, all results can by slightly overestimated.
46  The share of excluded commodity lines (CN 851710 and 851790) was close to 20 per cent of Polish imports in 1994.
47  Chris *et al.* (1998).

# References

Allen, Chris, Michael Gasiorek and Alasdair Smith (1998) 'The Competition Effects of the Single Market in Europe', Economic Policy 27, October, London: CEPR.
Armington, Paul (1969) 'A Theory of Demand for Products Distinguished by Place of Production', IMF Staff Papers xvi, p. 159–76.
Balcerowicz, Leszek (1995) *Socialism, Capitalism, Transformation*, Budapest: CEU Press.
Begg David (2000) 'Trade policy and macroeconomics in European transition economies', Paper prepared for conference on trade policies and economic transition, Chapter 3 above.
Eaton, Jonathan and Gene Grossman (1986) 'Optimal trade policy and industrial under oligopoly', *Quarterly Journal of Economics*, no. 101, p. 383–406.

*Europe Agreement Establishing an Association between the Republic of Poland, of the One Part, and the European Communities and Their Member States, of the Other Part* (1994) Dziennik Ustaw RP, zalacznik do numeru 11, poz. 38 z dnia 27 stycznia 1994.

European Bank for Reconstruction and Development (EBRD) (1999) *Transition Report.*

Francois, Joseph and Keith Hall (1997) Partial Equilibrium Modeling, in Joseph Francois and Kenneth Reinert (1998), *Applied Methods for Trade Policy Analysis,* Amsterdam: Erasmus University.

Kaliszuk, E. and E. Synowiec (1995) 'Klauzule Ochronne z Ukladu Europejskicgo: gdzie zastosowano i w czym pomolgy obu stronum, *Rynki Zagraniczne,* no. 24, 25 February.

Kawecka-Wyrzykowska, Elzbieta (2000) 'Lessons from trade liberalizaton in transition economies: ten years experience and future challenges', Warsaw: FTRI, Discussion Paper no. 87.

Kierzkowski, H., M. Okolski and S. Wellisz (eds) (1993) *Stabilization and Structural Adjustment in Poland,* London: Routledge.

Kostecki, Michael (1979) *East–West Trade and the GATT System,* Larter: Beutwer.

Krugman, Paul (1984) 'Import Protection as Export Promotion. International Competition in the Presence of Oligopoly and Economies of Scale', in H. Kierzkowski (ed.), *Monopolistic Competition and International Trade,* Oxford: Blackwell.

Krugman, Paul and Alasdair Smith (eds) (1994) *Empirical Studies of Strategic Trade Policy,* Chicago: University of Chicago Press.

Maliszewska, M., J. Michalek and A. Smith (1998) 'EU Accession and Poland's External Trade Policy', Economic Discussion Paper no. 45, Warsaw: Warsaw University Faculty of Economic Sciences.

Marczewski, Krzysztof (1999) *Szacunek efektów kreacji i substytucji handlu wyrobaprzemyslowymi wynikających z przyjecia wspólnej taryfy celnej Unii Europejskiej, w: Skutki przyjecia przez Polske wspólnej taryfy celnej Unii Europejskiej, 1999,* E. Kawecka-Wyrzykowska (red), Problemy handlu zagranicznego nr 21, IKCHZ, Warszawa.

Mayhew, Allan (1998) *Recreating Europe: The European Union's Policy Towards Central and Eastern Europe,* Cambridge: Cambridge University Press.

Orlowski, Witold (2000) *Koszty i korzyści z czlonkostwa Polski w Unii Europejskiej. Metody, modele, szacunki,* Warszawa: CASE.

*Poland's Foreign Trade Policy 1992–1993* (1992) Warsaw: Foreign Trade Research Institute.

*Poland's Foreign Trade Policy 1993–1994* (1994) Warsaw: Foreign Trade Research Institute.

*Poland's Foreign Trade Policy 1994–1995* (1995) Warsaw: Foreign Trade Research Institute.

*Poland's Foreign Trade Policy 1995–1996* (1996) Warsaw: Foreign Trade Research Institute.

*Poland's Foreign Trade Policy 1996–1997* (1997) Warsaw: Foreign Trade Research Institute.

*Poland's Foreign Trade Policy 1997–1998* (1998) Warsaw: Foreign Trade Research Institute.

*Poland's Foreign Trade Policy 1998–1999* (1999) Warsaw: Foreign Trade Research Institute.

*Raport w sprawie korzyści i kosztów intergacji Rzeczypospolitej Polskiej z Unią Europejską* (2000) Raport stanowi realizację uchwaly Sejmu Rzeczypospolitej Polskiej z dnia 18. lutego 2000.

Shiells, Clinton, Robert Stern and Alan Deardorff (1986) 'Estimates of the elasticities of substitution between imports and home goods for the United States', *Weltwirtschaftliches Archiv,* vol. 122, no. 3, p. 497–519.

Synowiec, Ewa (1995) 'Porównanie klauzul ochronnych w umowach Polski z WE, z krajami EFTA i krajami CEFTA, in *Umowy o wolnym handlu ze Wspólnotami Europejskimi oraz krajami EFTA i CEFTA i uzgodnienia Rundy Urugwajskiej; Skutki dla polskiegi handlu zagranicznego*, Warszawa: Instytut Koniunktur i Cen.

*Trade Policy Review Poland.* Report by the Government (2000) Geneva: World Trade Organization, document WT/TPR/G/71.

*Trade Policy Review Poland.* Report by the Secretariat (2000) Geneva: World Trade Organization, document WT/TPR/S/71.

*Umowy o wolnym handlu ze Wspólnotami Europejskimi oraz krajami EFTA i CEFTA oraz uzgodnienia Rundy Urugwajskiej: Skutki dla polskiego handlu zagranicznego*, (red.) Elżbieta Kawecka-Wyrzykowska, 1995, IKCHZ, Warszawa.

*Zagraniczna polityka gospodarcza i handel zagraniczny Polski 1999–2000* (2000) Warszawa: Instytut Koniunktur i Cen.

# 6
# Hungary: Stability of Trade Policy and its Determinants

*Sándor Meisel*

## The macroeconomic framework of Hungarian trade policy

Hungary was the first of the former CMEA countries to begin broad economic reforms. This started as early as 1968, but an active transition to a market economy did not begin until 1988. That year marked the establishment of the right of all companies and citizens to engage in trade within the convertible currency area. Legislation was adopted on foreign direct investment (FDI) guaranteeing investment protection and providing investment incentives. There was the introduction of comprehensive price and tax reforms, including the implementation of a VAT and personal income tax. Building on these and other measures, the newly elected government launched its programme for national renewal in 1990. The key objectives of this programme were macroeconomic stabilization and structural adjustment aimed at transforming Hungary into a market economy.

The programme included a fiscal austerity package designed to reduce a rising budget deficit. There were measures to contain Hungary's large and growing foreign debt, and a tight monetary policy to check inflation. Among the main structural features of the programme were greater currency convertibility, the further opening of the economy to international trade (*inter alia* through reduction in the scope of import licensing), and the intensification of links with Western Europe. This was accompanied by new legislation designed to facilitate FDI, a radical privatisation programme, and additional price and wage liberalization. Other reforms included major changes in company law, the establishment of a stock exchange and the adoption of a new competition law. Moreover, the reduction of subsidies and the process of deregulation were accelerated. These reforms were supplemented by social policies designed to alleviate the heavy unemployment initially associated with structural adjustment and to facilitate the restructuring of the Hungarian economy.

Following the collapse of trade between former CMEA members, Hungary's structural adjustment programme has seen a continuing shift in trade to industrialized countries, particularly EU members. Clearly, this reorientation has been greatly facilitated by the Europe Agreement and the creation of the CEFTA as well as by the WTO Agreements. Other important features of Hungary's structural adjustment have been the liquidation of a large number of companies owing to the implementation of a strict bankruptcy law. There has also been large-scale privatization of state-owned enterprises and a prominent role for foreign direct investment, not just in privatization but in the transition process in general. Privatization, together with reduced levels of subsidies, has greatly reduced the government's intrusiveness in economic decisions.

Among the principal macroeconomic developments at the beginning of the 1990s was a marked deterioration in overall economic performance. This was apparent as there was a notable 15 per cent decline in real GDP during the period 1991–93, as well as high inflation and rising unemployment. Another major development was the emergence of large and rapidly increasing budget and current account deficits. However, the introduction of a major stabilization package in March 1995 appears to have restored macroeconomic balance. This paved the way for sustained improvements in Hungary's economic performance.

On the basis of the macroeconomic developments in the second half of the 1990s one may conclude that the Hungarian economy seems to have reached the second stage of transformation. Annual growth is persistently high (4.5 to 5.5 per cent) and the officially registered unemployment rate is about 6 per cent and falling. The budget deficit is slightly decreasing and, more interestingly, the dynamic growth does not generate rapidly deteriorating trade and current account balances. Macroeconomic figures are based on a strong micro-economic performance dominated by international companies.

We can agree with Kaminsky (1999) that the rapid pace of the turnaround, after the dramatic contraction in exports in 1993, has had a great deal to do with both the emergence of 'second generation' firms and foreign participation. Foreign-owned firms tend to be more export oriented and more profitable than domestically owned firms. Thanks to a friendly environment for FDI since the outset of transition, Hungary has been the most successful transition economy in terms of attracting foreign investors. FDI has played a pivotal role in reintegrating the Hungarian economy into international markets. A huge portion of investment has come from large MNCs with global networks of production and marketing. As a result, a significant share of Hungary's domestic business activity has been incorporated into these networks. In addition, most FDI has come to Hungary, not only as a way of jumping trade barriers, but to take advantage of the overall economic environment (Kaminski 1999).

The main trade figures are shown in Table 6.1.

*Table 6.1* Hungary: indicators of Hungarian foreign trade (m USD)

|  | 1990 | 1991 | 1992 | 1993 | 1994 | 1995 | 1996 | 1997 | 1998 | 1999 | 2000[1] | 2001[2] |
|---|---|---|---|---|---|---|---|---|---|---|---|---|
| Exports | 9588 | 10187 | 10705 | 8907 | 10701 | 12867 | 15703 | 19100 | 23005 | 25013 | 28100 | 31500 |
| Imports | 8647 | 11382 | 11079 | 12530 | 14554 | 15466 | 18143 | 21234 | 25706 | 28008 | 31600 | 35000 |
| Trade balance | 941 | −1195 | −374 | −3623 | −3853 | −2599 | −2440 | −2134 | −2701 | −2995 | −3500 | −3500 |

*Source:* Central Statistical Office of Hungary.

*Notes:* 1 Including customs free zones
2 Forecast

Much like the general Hungarian macroeconomic policy, the exchange rate policy of the 1990s also can be divided into two phases. The most relevant and authentic description of Hungarian exchange rate policy during that period is presented by Szapáry and Jakab (1998). The following assessment is based on their article.

When inflationary forces are so dominant in a transition economy that they cannot be offset by productivity growth, and when a wage agreement cannot be struck because of the lack of policy credibility, the rate of inflation in that economy will also be higher than that of its trading partners. In that case, a depreciation of the nominal exchange rate is necessary in order to maintain competitiveness. Due to the fact that Hungary was burdened by a high level of external debt at the outset of the reforms, ensuring competitiveness and limiting the current account deficit were seen as sine qua non conditions for achieving sustainable economic growth. An exchange rate fixed to a strong currency, though, would have temporarily moderated inflation. However, it would have led to a loss of competitiveness and unsustainable current account deficits. Because the sustainability of such an exchange rate policy would not have been credible, it would not have reduced inflation expectations. In the 1980s, Hungary had experienced several episodes of severe balance of payments problems. Each time, the unavoidable adjustment involved a devaluation and a subsequent increase in inflation. As a result, whenever there was a serious deterioration in the external balance, inflation expectations were heightened. Given these past experiences, the moderating effect of a fixed exchange rate on inflation would have been short lived at best.

The adoption of a freely floating exchange rate regime was another option. However, it carried the risk that it would lead to large fluctuations in nominal and real exchange rates, particularly in light of the great uncertainties surrounding the effects of reforms, particularly those of the liberalization of trade and payments. This was considered to be unacceptable. The corporate sector could not have hedged itself against the potentially large exchange risks because the futures and forward markets were missing at the start of the reforms process and were developing only gradually. Therefore, the Hungarian authorities opted for an adjustable peg, under which the forint was adjusted from time to time taking into account developments in the balance of payments and inflation. The forint was fixed to a currency basket and was depreciated 22 times for a total of 87 per cent between January 1990 and February 1995 (see Table 6.2). The deprecations of the forint at frequent but irregular intervals were meant to safeguard competitiveness. The system, however, had several serious drawbacks. Firstly, a few weeks after each devaluation, the markets started to speculate about the timing of the next devaluation. This involved, inter alia, bringing forward imports and delaying the repatriation of export receipts. Such speculations distorted import and export statistics and made it difficult to assess the underlying trends.

*Table 6.2* Hungary: devaluations of the forint

| Date | Devaluation (%) | Date | Devaluation (%) |
|---|---|---|---|
| 31 January, 1990 | 1.0 | 9 July, 1993 | 3.0 |
| 6 February, 1990 | 2.0 | 29 September, 1993 | 4.5 |
| 20 February, 1990 | 2.0 | 3 January, 1994 | 1.0 |
| 7 January, 1991 | 15.0 | 16 February, 1994 | 2.6 |
| 8 November, 1991 | 5.8 | 13 May, 1994 | 1.0 |
| 16 March, 1992 | 1.9 | 10 June, 1994 | 1.2 |
| 24 June, 1992 | 1.6 | 5 August, 1994 | 8.0 |
| 9 November, 1992 | 1.9 | 11 October, 1994 | 1.1 |
| 12 February, 1993 | 1.9 | 29 November, 1994 | 1.0 |
| 26 March, 1993 | 2.9 | 3 January, 1995 | 1.4 |
| 7 June, 1993 | 1.9 | 14 February, 1995 | 2.0 |
| | | 13 March, 1995 | 9.0 |

*Source*: NBH, *Monthly Report*, 1997/10, cited by Szapáry and Jakab (1998).

Second, because policy makers in general are not too keen to devalue, each devaluation was preceded by an internal debate about the magnitude of the adjustment. The end result was that the cumulative devaluations were less than conditions called for, and the real effective exchange rate appreciated significantly, contributing to a worsening of the balance of payments.

In March 1995, a comprehensive stabilization-cum-reform programme was adopted. The main elements of the programme included expenditure cuts, increases in consumption, and value-added taxes. There was also the introduction of an 8 per cent temporary import surcharge, and a 9 per cent devaluation of the forint (see Table 6.3), followed by the introduction of a crawling band exchange rate system. At the same time, the pace of structural

*Table 6.3* Hungary: the preannounced monthly rate of the crawl of the forint

| | |
|---|---|
| From 16 March, 1995 | 1.6% |
| From 1 July, 1995 | 1.3% |
| From 1 January, 1996 | 1.2% |
| From 1 April, 1997 | 1.1% |
| From 15 August, 1997 | 1.0% |
| From 1 January, 1998 | 0.9% |
| From 15 June, 1998 | 0.8% |
| From 1 October, 1998 | 0.7% |
| From 1 June, 1999 | 0.5% |
| From 1 October, 1999 | 0.4% |
| From 1 April, 2000 | 0.3% |

*Source*: National Bank of Hungary.

reforms was stepped up. Capital transactions were further liberalized, privatization was accelerated, and various reforms in the areas of taxation and the social safety net, including a pension reform, were undertaken.

Table 6.4 shows that although the mixed priorities of exchange rate policy have been rather stable in the 1990s their ranking was modified from time to time and adjusted according to the internal macroeconomic situation and to the needs of easing of the external pressure.

Despite a number of temporary economic difficulties and some cases of economic policy shortcomings, one has reason to consider the economic performance of Hungary during the 1990s as a successful transition.

There have been a number of factors which led to such a favourable setting. Hungary had started to establish the framework for a market economy well before 1989 (legal and institutional reforms, cooperative links with Western firms, a more outward-looking society, and so on). Perhaps they were driven less by conviction and strategic thinking than by the growing pressure of external indebtedness. The country, however, opted for a privatization programme conforming to market structures and based on opening up to foreign capital. In the 1990s, Hungary became one of the most internationalized economies by global comparison, and is an essential location for multinational production (and services) to the European and global markets (Inotai 2001). Certainly, the Association Agreement signed with the European Union and the subsequent liberalization of foreign trade offered a chance to reorient trade relations within a very short period. This has led to their acquisition of new markets and the development of a dramatically different and competitive manufacturing structure. To put it in a simplified way, Hungary's economic success is the result of twin developments. First, the new demand that opening markets generated, but also the new supply that the rapidly growing internationalization created within a very short period.

*Table 6.4*  Hungary: priorities of exchange rate policy

| 1990 | Structural change, competitiveness |
|------|-------------------------------------|
| 1991 | External balance, anti-inflation, competitiveness |
| 1992 | Anti-inflation |
| 1993 | Anti-inflation, maintaining of the real exchange rate |
| 1994 | Competitiveness of exports, anti-inflation |
| 1995 | Anti-inflation, improvement of external balance |
| 1996 | Anti-inflation, competitiveness |
| 1997 | Anti-inflation |
| 1998 | Anti-inflation, improvement of credibility |
| 1999 | Sustainable economic growth |

*Source*:  Erdős and Molnár (1999).

The interrelationship between *trade policy* and *macroeconomic policy* has been rather intimate. The three-year programme of opening up the economy between 1989 and 1991 was meant to break the backbone of the socialist dinosaurs. Maintaining an open trade regime was seen as a necessary precondition for structural changes, which actually materialized. Last, but not least, it was thought that in a small country with lots of large economic units, the monopolistic practices of FDI could only be counterbalanced by an open trade regime. This way competitors will take care of what no one else could. The latter actually materialized, with large investors often using the otherwise dormant stipulations of the competition law against each other (Csaba 1996).

## Development of the main trade policy elements

This section of the study focuses on the most important developments of Hungarian trade policy during the 1990s. The history of the trade policy through this period can be divided into two different stages. The first stage of this evolution in policy commenced at the beginning of the decade (more precisely, from 1989 and 1992). The second stage was longer and lasted from 1992 (starting the implementation of the Europe Agreement) until 1 January, 2001. This stage concluded with the accomplishment of the transition period of the Europe Agreement (and also most of the other free trade agreements). Theoretically, a third stage can also be distinguished in this history. This lasted from 1 January, 2001, until the accession of Hungary into the European Union. More precisely, until the end of transition measures of the Accession Treaty. This is the last period of an autonomous Hungarian trade policy. Our analysis concentrates mainly on the first two stages and the concluding section will briefly elaborate on the accession conditions of the third stage.

The basis for the logical destination of the first and the second phases consists in the different frameworks of the liberalization. While, in the first stage, a unique multilateral approach to liberalization dominated, conversely, the second stage is characterized by liberalization steps taken mainly within the framework and according to the provisions of bilateral free trade agreements. This was complemented by liberalization commitments taken in the GATT and WTO framework. Nevertheless, the most important feature of these two phases of Hungarian trade policy is general liberalization.

### First stage of trade policy liberalization (1989–92)

Before the conclusion of the Europe Agreement and similar free trade agreements with Western and Central European countries, Hungary had only one free trade agreement. This was the agreement with Finland on the Removal of Obstacles to Trade which was concluded in 1975. The gradual removal of tariffs was implemented by 1 January, 1985. In addition to industrial products, the agreement covered certain agricultural goods.

The most important liberalization measure in this period was the elimination of the bulk of quantitative restrictions. This was done on an *erga omnes* basis. The unilateral process started in 1989 and was accomplished in three years. It should be mentioned that within the Hungarian trade policy regime of the 1980s, these restrictions played a dominant role in the import policy. However, the effect on the customs regime was only secondary. Quantitative restrictions covered almost 100 per cent of Hungarian imports before liberalization. As a consequence of this process, their coverage diminished in three years to about 15 per cent. This decision was motivated by the logic of liberalization of the economic system. This was part of the process of transition to a market economy. After the conclusion of free trade agreements, entering into force in 1992, the liberalization of the quantitative restrictions (both individual licensing and the global quota for consumer products) changed. This was the same for the bilateral agreements and the remaining trading partners. The free trade agreements contained a special timetable for the elimination of the QRs, giving growing preference to such partners.

Although falling in the period between 1989 and 1992, the quantitative liberalization was based on the *erga omnes* principle. This easing of the Hungarian import system appeared to be beneficial mainly for the exporters of the EC member states. This was due to their geographical location, traditional links, and market experience, along with Hungarian consumers' preferences. This is shown by the increase of the EC's share in Hungarian imports in that period.

Certain other developments in the Hungarian trade policy regime should also be mentioned. These are already related to the initiation of the preparatory works of the European Agreements.

One such matter is connected with a specific feature of Hungarian trade policy. In the process of meeting the country's obligations in the Tokyo Round, Hungary made some discriminatory distinctions between the EC and other trading partners. Hungary, in the case of trade with the Community member states, unilaterally postponed the application for the Tokyo Round tariff reduction for certain products. This was a countermeasure against the EC's maintenance of some specific quantitative restrictions against Hungarian imports. When the EC eliminated these restrictions by 1990, Hungary returned to the original Tokyo Round tariff reduction scheme in trade with the EC the next year. The basic duties of the Europe Agreement thus were established according to these original commitments concerning tariff reduction.

The other matter related to the fixing of the basic duties in the framework of the Europe Agreement was the correction of tariffs within certain product groups. Hungary, in 1990 and 1991, notified GATT of increases of bound duties for passenger cars, colour TV sets, and some basic chemical materials.

The reasoning behind this measure was based on several considerations. One of them was that Hungarian trade policy was parallel to the elimination of the quantitative restrictions and therefore tried to increase the effect of duty protection for these products. The other intention was to provide increased protection for production facilities which were newly established in the country (e.g. GM and Suzuki) and for products produced by industries intending to open new plants in Hungary (e.g. Samsung). Nevertheless, it should be noted that the product coverage of such Hungarian initiatives was and remained limited. This can be explained by the fact that a dominant part of the industrial duties was bound in the GATT. Hungarian authorities were restrained in starting a doubtful process of notification and compensation.

Some important institutional and legislative developments took place in this first phase of Hungarian trade policy, as part of the creation of market economy institutions and procedures. The laws were created to regulate protection in the market. The circumstances in which a safeguard clause may be used are stipulated in Government Decree no. 113/1990 on market protection measures. The terms of this followed Hungary's protocol of accession to GATT. This was defined as a special, selective protection clause. Dumping proceedings in Hungary were originally governed by Government Decree no. 111/1990, suspended later by no. 69/1994. This original legislation reflected the results of the Tokyo Round of GATT and was contrary to the EU's regulation. The EU regulation was based on the results of the Uruguay Round and, therefore, there was some disharmony. The conflict was resolved in 1998, when the results of the Uruguay Round were embodied in Hungarian law.

### Second stage of trade policy liberalization (1992–2001)

As it was mentioned before, this second phase of trade policy can be distinguished from the first one by a strong shift from multilateral measures to regional and bilateral processes. Although Hungary sees multilateral and regional liberalization as complementary processes, regional liberalization has indeed proved to be a major element in Hungary's opening up and transformation. The lion's share of the measures aiming at trade policy liberalization was taken in the regional context and in the framework of bilateral free trade agreements. It does not mean neglecting trade policy liberalization on the basis of GATT and especially WTO commitments but, nevertheless, the bilateral provisions served the driving force in this process. Certainly one may evoke some criticism in this respect (see, for example, Messerlin 1998, Winters 1993), but this remains an unavoidable fact.

Another important feature of this process is that liberalization covers all traditional trade policy instruments and fields. They sometimes also went beyond these traditional measures. The European Agreement provides for the elimination of all classical trade policy barriers. In many respects the other free trade agreements go in the same direction. The process of EU accession gradually involves a growing number of regulations for the legislative harmonization. This leads to the shrinking of space for applying restrictive measures to be able to affect trade. This is notable in the relationships between free trading partners.

There is also an original, interesting, and to some extent neglected feature of the bilateral trade policy liberalization. In quite a number of cases, provisions aiming at bilateral trade liberalization became multilateral. There are several examples, when commitments originating from one or more free trade agreements have been extended to trading partners. This is applicable to all WTO countries outside the agreements concerned, on an *erga omnes* basis. On 1 January, 1997, Hungary eliminated all charges, which had an effect equivalent on customs duties on imports from the EU and other free trading partners. At the same time, this measure was extended to all WTO countries. Since 1998, Hungary has not applied quantitative restrictions on textile products against WTO members. This was also stipulated by the Europe Agreement and other free trade agreements. The same happened in 2001, when Hungary abolished quantitative restrictions in the form of the global quota on consumer products not only in trade with free trading partners but also in relation to all MFN partners.

Among the free trade agreements concluded by Hungary in the 1990s, the Europe Agreement had the greatest influence. This is because it was signed before the other treaties and it covers the largest part of the country's trade. It is also because it served as a model for all the other free trade agreements, which reflect the logic of liberalization of the EU association agreement. The exception was the principle of asymmetry, which was included only in the Europe Agreement and the EFTA free trade agreements.

Thus, the most important impact on the liberalization of Hungarian trade policy comes from the agreement signed with the EU and the member states parallel to the other – at that time – two so-called Visegrád countries. The logic, scheme and even the timetables of liberalization of trade are generally very similar in all of the first generation Europe Agreements.

The asymmetrical establishment of the free trade area in the industrial sector and the asymmetry in agricultural trade concessions correspond to the original Hungarian intention. According to it, the EU abolished all quantitative restrictions, except those in the textile products trade. The customs duties on industrial products were eliminated on the EU's side according to the fixed timetables. They were accelerated after the well-known decision of the 1993 Copenhagen European Council Summit.

The process of Hungarian duty elimination was divided into three parts. From inception until 1 January, 1994, Hungary eliminated the duties of the so-called 'quick list' in three equal phases. The share of the concerned products of dutiable industrial imports was about 15 per cent at the date of the EA's signing. Imports of goods listed here were marginal both from fiscal and structural points of view. The criteria of being placed on this list were a relatively low level of duties and the minimization of economic effects.

Machinery and chemical products, consumer goods, metal and metallic products were mainly listed in the second, so-called 'normal list'. The duties on these products were eliminated between 1995 and 1997 in three steps. The share of these products in the Hungarian industrial imports was about 20–25 per cent in the period preceding the association. It is interesting to note that in this period of the duty elimination, the share of this group increased to about 50 per cent. This reflects the structural changes in Hungarian imports from the EU. However, trade in these products was almost entirely liberalized from the quantitative restriction point of view. The increase in the share of imports was mainly due to this fact and not to the provisions of the Agreement.

The structure of the third and so-called 'slow list' consisted of products for which duties were eliminated relatively slowly during the period between 1995 and the end of 2000. This was similar to that of the normal list, except for textile and steel products (those subject to gradual duty elimination or existing quantitative restrictions on the EU's side), most of which have been listed here.

It is interesting to note that among the original Hungarian ideas was the aim to form a customs union with the EU. Some of the earlier association agreements signed by the EC included provisions on the creation of the customs union. The EU flatly rejected this proposal. This could have been a symbolic element, but its economic impact was never analysed before the preparatory works of the accession.

In conclusion, the elimination of duties in the trade of industrial products between the EU and Hungary has been accomplished according to the original timetable. There was some backsliding, though, which will be analysed in the following sections.

Returning to quantitative restrictions on industrial products in Hungarian – EU trade, Hungary respected the obligations of the European Agreement and decreased the coverage of the remaining individual licensing in this trade to 40 per cent by the end of 1997. By 1 January, 2001, all individual import-licensing schemes were abolished. Concerning the other forms of quantitative restriction, the global quota on consumer products and the EU's subquotas were handled according to the provisions of the European Agreements. All kinds of similar restrictions were abolished by the same date.

The Europe Agreement contains limited concessions for the agricultural trade. This agreement was designed on a product by product basis. The original scheme was extended after the accession to the EU of certain EFTA countries. It was modified to take into consideration the results of the WTO agreement. After July 2000, these concessions were further extended following a new logic, one that provides for mutual elimination of export subsidies for certain products, and since then a new scheme has been set up.

In March 1993, Hungary signed free trade agreements with countries of the EFTA. This agreement was modelled according to the structure, content and timetables of the Europe Agreement. The aim was to establish identical or similar horizontal rules across free trade agreements in Western Europe, albeit with some differences in services and in agriculture. For Hungary, liberalization schedules for industrial products in the Europe Agreement and the EFTA are identical. Duties and quantitative restrictions on Hungarian industrial exports were eliminated by 1 January, 1997; Hungary removed all such measures on 1 January, 2001.

In the framework of the CEFTA agreement (with Hungary as one of the founding members), all duties and quantitative restrictions were gradually and systematically eliminated by 1 January, 2001. Liberalization of trade in agricultural and food products is also in an advanced stage. This is even larger and more important than the original Europe Agreement scheme.

To complement its network of European free trade agreements, Hungary concluded similar treaties with Bulgaria, the three Baltic states, Turkey and Israel between 1998–99. These agreements provided for the phasing out of industrial tariffs and their elimination by 2001. Trade in agricultural products is subject to various tariff concessions. Also of note, a free trade agreement with Croatia was concluded in 2000.

These developments greatly affected Hungary's customs policy during the 1990s. Nevertheless, one should not forget the Hungarian commitments taken in the multilateral framework. According to the latest WTO Trade Policy Review of Hungary (1998) the coverage of tariffs increased to 95.7 per cent of total tariff lines in 1997. This was partly as a result of Hungary's WTO commitments under the Uruguay Round. The average applied MFN tariff rate has risen from 11 per cent at the beginning of the 1990s to 14.3 per cent in 1997. This was mainly a consequence of the tarification of agricultural non-tariff measures, the introduction of the Harmonized System (HS) in 1992, and the EU's Combined Nomenclature (CN) in 1996. The scope of Hungary's preferential agreements with the EU and other countries is such that MFN tariffs apply in practice to about 20 per cent of Hungary's imports. As a consequence, the collected rate of duty on all imports was about 2–3 per cent in the last years of the decade. More than 500 of the tariff rates that were applied in 1997 were less than the final bound Uruguay Round rates. If these tariffs remained at the same rate until full implementation, the overall applied MFN rate would decline from 14.3 to 9.6 per cent.

It should be mentioned here that Hungary had been operating a generalized system of preferences (GSP) on imports from developing countries since 1972. The system has been applied on a permanent basis and does not involve time limits or renewal requirements. At the end of the 1990s, GSP tariffs were applied to about 6400 tariff lines. Out of these tarriffs, only 40 per cent were of zero rate and the rest were 20–90 per cent lower than the MFN rate. The number of beneficiary countries is at present about 110. This regime covers about 95 per cent of total imports from developing countries.

While analysing Hungary's customs policy, one should not forget an original and specific feature of this policy. This is the issue of customs free zones. This not an unknown practice for Central and Eastern Europe or for the EU. However, the scope and importance of this practice, due to some specific Hungarian regulations, distinguishes this country from many others. At the end of the 1990s, there were more than 100 such zones and more than 70 companies had been established there, mainly multinationals. The role of the customs free zones in the Hungarian economy can be reflected by their foreign trade activity. The share of Hungarian exports in such zones increased between 1996 and 1999 from 18 to 43 per cent. The figures for imports are 16 and 30 per cent. Among the 10 largest Hungarian exporting firms, 8 have been established in customs free zones (such as Audi, IBM, Philips, Opel, Flextronics and so on). The main reason for being established in such zones is the specificity of Hungarian regulation. It permits a dispersed geographical location across the country. More importantly, in Hungary, equipment and investment goods used for manufacturing are allowed to enter the zones customs free. These are the main differences in comparison to the EU regulation and it may explain the striking advantage of this customs construction. On the other hand, it may be problematic in the process of the taking over of the Common Customs Policy.

As part of the customs liberalization of the 1990s, Hungary abolished all charges on imports that had an effect equivalent to customs duties. These were a licensing fee of 1 per cent, customs clearance fee of 2 per cent, and a statistical fee of 3 per cent. Their level might have been considered rather high by international comparison. These fees were gradually eliminated according to free trade agreements between 1995 and 1997, and the practice was extended to all WTO countries.

As mentioned before, one of the most important liberalization measures that took place during the 1990s liberalization of the Hungarian quantitative import regime. Two elements of this system, individual licensing and global quota on consumer products, have been greatly affected. They played a decisive role in the trade policy of the previous decades but in the 1990s, due to their gradual removal, they largely lost their functionality.

In the 1990s, individual licensing systems covered such imported products as pharmaceuticals, telecom equipment, coal, iron ore, certain chemicals, agricultural wood products. According to the free trade agreements, these restrictions have been gradually abolished and this elimination has been extended to the WTO countries. At present import licences of this type are only applied to goods that are typically controlled in most countries (products for military purposes, hazardous materials, and so on).

The other element of the quantitative regime, the global import quota on consumer goods, was a specific Hungarian feature. It was introduced in the 1960s and was based on balance of payments considerations. It contained subquotas broken down according to certain product groups(detergents, footwear, textiles and so on and, according to the Europe Agreement, also passenger cars). After the preferential agreements came into force, the subquotas were established according to groups of countries (e.g. EU and CEFTA) The subquotas were used as ceilings and, in many cases, licences were issued over the quotas. From 1 January, 2001, the global quota on consumer goods was abolished in relation to both preferential partners and all WTO countries.

A monitoring system on certain imports was set up in 1993. Under the monitoring system in the 1990s, licensing was automatic but was liable to administrative procedures.

As mentioned before, starting from the early years of the 1990s, Hungary had established and operated its safeguard legislation. In that period, Hungary frequently used safeguard measures as a tool of market protection. This is mainly due to the fact that this tool was not very difficult to introduce. In most of the cases, it was used against imports coming from countries which were not members of the WTO. In the early 2000s, such cases have become less frequent. Nevertheless, this instrument can be considered as the most powerful, or, at least, most unique, market protection instrument used during the 1990s. Detailed cases of its application are described in the following section. According to the Europe Agreement (Art. 28), Hungary had the option to use a unilateral tool to protect its infant industries or those under restructuring in the form of temporary duty increase.

Although Hungary does have legislation concerning anti-dumping and countervailing measures (based on Government Decree no. 69/1994), these measures have never been used. Very few requests have been made for anti-dumping investigations and the authorities have always refused them. No request has been made for countervailing investigations. It seems that the procedure required for such an investigation is too complicated for domestic producers and is probably too costly.

Among internal measures having a potential impact on trade are quality, labelling and safety requirements. In addition, consumer protection laws, environmental standards and sanitary measures can be mentioned as potential tools for influencing imports. The room for the application of these measures is shrinking as the legal harmonization, including the taking over of the EU's internal market regulation, has greatly advanced. On the other hand, in a period when the role of classical trade policy instruments in shaping the pattern of trade is diminishing, measures and instruments falling into this category are becoming more important. Hungary, during the 1990s, had developed a large system of such legislation, which was clear and transparent, at least on the level of legislation. Nevertheless, on the level of the application and functioning, it appears that this system is less sophisticated and is not always equipped with an appropriate infrastructure, as compared with, say, the EU. The next section will provide examples of using these tools.

Finally, regulation of exports also should be mentioned briefly. Currently less than 2 per cent of Hungary's national tariff lines are subject to export licensing. This system was applied in the 1990s in some sectors where Hungary had to respect its international obligations (e.g. textiles). Nowadays, it covers mainly environmentally sensitive products, which is similar to the practice of most other countries. There were also several cases in the 1990s, when Hungary applied this tool of trade restriction on the basis of real or possible market shortages, especially in the grain sector.

## Slippages of Hungarian trade policy

This section contains an overview of the Hungarian trade policy measures taken during the 1990s. During this period, there were some deviations from the general rule, tendency and logic of the liberalization process (Table 6.5). In most cases, the application of these measures produced nothing special. They are frequently used in international trade policy practices. Nevertheless, they reflect a situation when the existing trade policy regulation is considered to be not protective enough or when specific public policy and sectoral considerations prevail over the general tendency.

The analysis of this section follows the structure of the measures reflected in the tables presented in this section.

The most important general trade policy deviation from the prevailing logic, which had an across-the-board effect, was the introduction in March 1995 of an import surcharge of 8 per cent. This was implemented on the *erga omnes* basis (see Table 6.6). It was one of the main features of the March 1995 package of stabilization measures. These were designed to address serious macroeconomic imbalances. Its temporary character was stressed from the introduction. This measure was notified to the WTO and justified by

*Table 6.5* Measures taken on the basis of national customs legislation

| Legal basis | Application | Content | Countries effected |
|---|---|---|---|
| Government Decree no. 21/1976 | 1992 | Tariff preference (0% duty) for motor vehicles falling under a newly established specification (tailored to Ford Transit) | *Erga omnes* |
| | 1994–1995 | Increase of duties (from 8–40°C to 40–70%) for certain agricultural products with unbound rates in the GATT | *Erga ommes* |

*Table 6.6*  Balance of payments measures

| Legal basis | Application | Content | Countries effected |
|---|---|---|---|
| Art. 63 of Europe Agreement, Art. 32 of CEFTA, Art. 24 of EFTA, notified to the WTO | 1995–1997 | Import surcharge of 8% subject to gradual elimination | *Erga ommes* |

the Hungarian government on balance of payments grounds. The surcharge was applicable to imports from all sources and covered all products except primary energy products. It was refundable in the case of machinery imported for investment purposes. The surcharge was included in the selling price upon which VAT was levied. The overall restrictive effect of the surcharge considerably exceeded 8 per cent. In the case of duty free import products which were subject to the standard VAT rate of 25 per cent, the effective rate of the surcharge was 10 per cent and not 8 per cent. The effective rate was even higher for products subject to import duties as well as excise taxes on top of VAT. It was gradually reduced to 6 per cent on 1 October, 1996, 4 per cent on 10 March, 1997, and finally to 3 per cent on 15 May, 1997 before being eliminated on 1 July, 1997.

As far as the measures taken on the basis of national customs legislation are concerned, one such measure was introduced in 1992. This was just after the implementation of the Europe Agreement. At that time, Ford was considering making important investments in Hungary. As part of the incentive package, besides tax and other preferences allowed by the legislation of that time, trade preferences were also shaped. This meant that customs tariff specifications for vehicles in the Ford Transit category were modified. This allowed the vehicle to enjoy duty free market access and thus had an important preference as compared to the competitors. In a very short time, this measure was sharply contested by the European Commission because of discrimination, which was incompatible with the Europe Agreement. The Hungarian authorities could not re-establish the original duty because it would have contradicted the standstill provision of the EA. The 0 per cent duty therefore remained in force for all types of the vehicles of this category. (It is interesting to mention that the customs authorities later led an investigation to establish the number of Ford Transit cars that entered duty free. It appeared that besides this type, many other cars of this category underwent the customs clearance procedure with 0 per cent duty. This reflects the administrative shortcomings of the customs authorities of that time.)

The other measure of this kind was the increase of certain non-GATT agricultural duties. At that time, considerable criticism was formulated concerning the agricultural trade scheme of the EA and the sudden increase of agricultural imports from the EU associated with the impact of the Agreement. These measures meant an early application of the tarificated duties in most cases. As a result, protection of the Hungarian agricultural market had already begun to strengthen by November 1994, instead of 1 January, 1995 (the date of the implementation of the WTO agreement). Nevertheless, besides the market protection effect, this step may be considered as a symbolic gesture in favour of agriculture. This is reflected by the fact that tropical products were also included in this regulation. Another motivation aside from mere gesturing might have been budgetary considerations to increase customs revenue.

Article 28 of the Europe Agreement – much as similar treaties with other Central European countries – contained a unilateral provision in favour of the associated states taking exceptional measures to afford temporary protection to infant industries and those undergoing restructuring. The timetable and conditions of such measures have been fixed in the EA. Shortly after the implementation of the Interim Agreement, in February 1992, Hungary formulated a request to use this tool in the case of 16 products. Amongst them were passenger cars that could have been considered as subject to the infant industry clause. After protracted consultations, Hungary withdrew many products from the request list and the EU accepted the justification for the temporary increase of duties in the case of eight products, belonging to three product groups (see Table 6.7). Therefore, Hungary was able to benefit from this possibility only to a limited extent, especially as compared to some other associated countries.

As mentioned previously, Hungary maintains a system of import monitoring as a preliminary element of its market protection system. In many cases, products become subject to monitoring for which the application of

*Table 6.7* Measures taken on the basis of the infant industry and restructuring clause

| Legal basis | Application | Content | Countries affected |
| --- | --- | --- | --- |
| Art 28. of Europe Agreement | 1995–2000 | Increase of duties for insecticides, fungicides, herbicides, disinfectants | EU member states |
| | 1995–2000 | Increase of duties for wood-free paper and coated paper | EU member states |
| | 1995–1997 | Increase of duties for tempered glass and laminated glass | EU member states |

*Table 6.8* Monitoring

| Legal basis | Application | Content | Countries affected |
|---|---|---|---|
| Decree no. 24/1993 of the Ministry of International Economic Relations | 1994–1996 | Monitoring of imports of tiles | Italy, Spain |
| | 1995– | Monitoring of imports of washing machines | *Erga omnes* |
| | 1995– | Monitoring of imports of milk powder, eggs, poultry | *Erga omnes* |
| | 1998– | Monitoring of imports of pigmeat | *Erga omnes* |
| | 1998– | Monitoring of imports of asbestos | Kazakhstan |

the safeguard clause was requested but the evidence of market injury could not be established (see Table 6.8).

The general safeguard clause is the most frequently used market protection tool of the Hungarian import regime. This fact is reflected in Table 6.9. Hungarian authorities introduced erga onmes based safeguard measures in

*Table 6.9* General safeguard measures

| Legal basis | Application | Content | Countries affected |
|---|---|---|---|
| Government Decree no. 113/1993 Art. 30 EA Art. 28 CEFTA Art. 20 EFTA | 1992–1993 | Import restriction on cement Import restriction on intra-ocular lenses Import restriction on certain paper products | *Erga omnes* |
| | 1995–1996 | Quota on imports of cement | Rumania |
| | 1998–2002 | Quota on steel products | Russia, Ukraine |
| Government decree no.113/1993 | 1998–2000 | Quota on certain steel products | Ukraine |
| | 2000–2004 | Quota on steel products | Russia |
| | 1999–2003 | Surcharge on imports of ammonium fertiliser | Russia, Ukraine |

only three cases and this happened before the implementation of the WTO agreement. After 1995 such measures have been taken in trade with Eastern European countries, Republic of the CIS, and countries that are not members of the WTO. Utilisation of this instrument is seemingly concentrated in a few industries. Application of the general safeguard clause by Hungary was most frequent in the steel sector, followed by the cement industry including products made from asbestos.

This fact deserves some elaboration. Let us turn first to the steel sector, the production of which, to a great extent, was affected by structural changes for the period of transition to a market economy. In the early 1990s, the representative associations of steel producers took a pro-liberalization attitude. As a result, practically all the quantitative restrictions were eliminated for the imports of steel by 1991. The collapse of the traditional CMEA markets resulted in market difficulties for the traditional productions (machinery, buses and so on) that were Comecon oriented. This meant that the steel producers lost their traditional domestic purchasers. Besides this, important structural changes took place in the domestic market, which was independent from the collapse of the Comecon. Thus, the internal demand for steel products diminished considerably (due to restructuring or poor economic position of the steel user industries and companies, e.g. the construction industry, manufacture, State Railway Company) At the same time, due to the supply of steel products from the overdeveloped production capacities of neighbouring countries, (Slovakia, partly Poland, the Czech Republic and the republics of the former Soviet Union) Hungarian steel producers, especially in the north-eastern part of the country, had to face strong competition in the narrowing domestic market. The imported products frequently came from producers functioning under distorted, and highly regulated conditions. In this situation, the safeguard clause became the only efficient tool for protection. The position of the industry was complicated by the fact that no clear industrial policy was formulated to restructure the Hungarian steel and iron sector.

The situation in the cement industry was a little different, although there were some similarities. In this case, the producers had to face competition from cheap exports of some neighbours with highly imperfect markets. The difference is that this industry has been completely bought out by foreign (German, Swiss) companies that have strong bargaining power in their relationship with the authorities.

Besides the aforementioned slippages of Hungarian trade policy, other slippages should also be mentioned. However, due to their specificity, they are not included in the tables in this section.

In the 1990s, Hungary, as almost all the European trading partners, used import restrictions several times on the basis of phytosanitary and veterinary regulations. From time to time, Hungary applied such restrictions vis-à-vis the EU members and CEFTA countries.

When speaking of internal measures affecting imports, an important step should be mentioned. Under the Customs Duty Law of 1995, imports of cars more than four years old were prohibited between 1995 and 2000, on environmental and safety grounds. An exception involved specialized older vehicles, which were allowed to be imported, provided they passed a special technical test. Along with high tariffs on cars imported from MFN partners, this measure benefited domestic car manufacturers and also EU and CEFTA producers enjoying growing tariff concessions.

Turning to the export regulations, one can observe that some backsliding took place. This is entirely within the agricultural trade sector. During the 1990s, Hungary applied for temporary export restrictions on certain agricultural products (mainly animal feeds, wheat and sweet corn) several times on the basis of the shortage clause. Since the statistical and information system of Hungarian agriculture often seems far from transparent, it is difficult to judge the justification of such measures.

Another well-known and major slippage of trade policy was also in the agricultural sector and relates to exports, namely to export subsidies. In accordance to its obligations under the Agreement on Agriculture, Hungary was committed to reducing agricultural export subsidies by 36 per cent between 1995 and 2000. This commitment initially involved a base amount which was set at Ft 22 billion. This implied a cut in the total amount of such subsidies to Ft 14 billion by 2000. Shortly after the implementation of the Uruguay Round Agreements, it became evident that the export subsidies in both 1995 and 1996 were about twice the level of Hungary's commitment of Ft 22 billion in nominal terms. It also covered 149 products instead of the 16 specified in the WTO Schedule. This situation was due to a considerable discrepancy between the value of total export subsidization in the period between 1986 and 1990, which constituted the base level for the reduction of subsidies, and the way it was reflected in the Schedule. The outcome was a dispute with several WTO members. The dispute was resolved in October 1997, with Hungary being granted a waiver from its obligations until 31 December, 2001. Under the waiver, the base amount was set at FT 51 billion, with Hungary agreeing to cut the amount to Ft 42 billion in 1997, then in stages so as to reach Ft 22 billion by 1 January, 2001.

## Determinants of trade policy

It is not easy to point to a few determining elements that shaped Hungarian trade policy during the 1990s. Determinants sometimes are hidden and, in many cases, their impact is not direct but is often interrelated. Therefore, the following description is merely an attempt to discover some possible elements that played a significant role in Hungarian trade policy formulation.

Probably, the most important determinant of the Hungarian trade policy during the 1990s was a strong commitment to achieve a successful transition to market economy. In addition, they wanted to advance its integration into the world economy and establish close links with the European Union, while taking into consideration the accession to the EU as a strategic objective. These commitments may explain the fact that the strategic and dominant tendencies of economic and trade policy liberalization, despite some slippages, has been maintained. Thus, the institutional and legal instruments, both GATT and WTO objectives and preferential agreements, maintained this general track. Certainly, in some cases and periods, especially in the early years of the 1990s, there were notable discrepancies. This was between the pace of liberalization optimal from the point of view of the internal transformation and that prescribed by external commitments. It is also true that the external agreements limit the room for autonomous actions. Nevertheless, this external institutional setting served as an important stabilizing element for liberalization. The initiation of accession negotiations even reinforced this impact.

Obviously trade policy can hardly be shaped and implemented without certain interactions with other internal policies. The direct link between trade policy and major backsliding in Hungary during the 1990s is evident in the case of the macroeconomic stabilization programme and the introduction of the overall import surcharge. A part of this measure, the macroeconomic policies did not directly affect trade policy formulation. At least, they did not enforce any other deviation from the general line.

Referring to the slippages of trade policy analysed in the previous section, a clear determining impact of financial and budgetary considerations can be observed in the case of the increase in agricultural duties. It can also be observed in the early application of the tarificated level of duties to some products in 1994.

Developments and backsliding of trade policy on sectoral level can be explained differently.

In Hungary, most of the protective measures (in the form of safeguards) were taken in the steel sector. Some explanation for this was given earlier. Here, it seems, the negative consequences of premature liberalization of the quantitative regime, the lack of a clear industrial and regional development policy, and the resulting employment problems were compensated by increasing the level of trade policy protection.

This is partly true for the cement industry, where the actor's bargaining power is much stronger, although the economic situation was similar to that in steel production.

In the paper industry, which also enjoyed market protection, a strong foreign company acquired the most important production plants. Shortly after this, the company was able to prove the potential injury of imports which then led to the application of a safeguard measure.

These two cases give some evidence for what was predicted in early 1990s. This was that a more powerful foreign company is better situated to request and argue for protective measures than traditional domestic firms. It should be noted that this situation has somewhat changed as domestic companies have also learned how to protect their interests.

A rather coherent picture can be observed in the car industry. At the beginning of the 1990s, this was an entirely new industry in Hungary. During the association talks, Hungary was able to resist the strong pressure of the EU aimed at a rather fast liberalization of this sector. According to the provisions of the Europe Agreement, trade of passenger cars was to be liberalized only at the last stage. A rather strong tariff protection is maintained in relation with MFN countries, which is often subject to criticism, mainly by the USA. This high level of protection was complemented between 1995 and 2000 by a ban on the importing of cars more than four years old.

It is interesting to note that there are traditional sectors that remained relatively less protected, although their professional interest groups had consistently argued for stronger protection. The pharmaceutical industry is one such example. One should not forget that the trade of pharmaceuticals is not a pure trade policy issue and is closely related to the social security system. Until 2001, this trade was subject to individual licensing, yet industry representatives always complained of permanent overlicensing.

As in most of the European countries, including those of the EU, trade policy regulation of the agricultural sector is a specific issue. Trade policy formulation in this sector is highly influenced by internal political considerations. This is true of Hungary, as it is in many Central and Western European states. Internal trade policy disputes are immediately transmitted on the highest political level. That makes agricultural trade policy rather unstable and vulnerable. But this is only one element of the instability.

The most important element of this instability is the lack of a well-defined agricultural policy. It is rather difficult to build up a consistent agricultural trade policy without the basic policy. However, interpreting the matter in a simplified way by looking at the system of information concerning the agricultural sector (who is producing, what is produced, how much is produced) can only be established with difficulty. It is hardly possible to formulate a coherent trade policy on this unstable basis. From a political point of view, this is reflected in symbolic measures (like the tariff increase in 1994), or in sudden and questionable restrictions on exports. Contrary to some other Central and Western European countries, the level of political sensitivity and economic uncertainty are the main determinants of the agricultural trade policy and not the simple existence of a strong agricultural lobby. The agricultural society in Hungary is dispersed and there is no union interest group which is able to represent the majority of the sector. Nevertheless, it is important to note that this sector is not responsible for the highest level of trade policy slippages.

## Future trade policy challenges and dilemmas

Until their accession to the EU, Hungary and the other candidate countries of Central and Eastern Europe preserved a certain amount of room to shape their own trade policy and to make, if necessary, the adjustments facilitating the future integration in the EU's customs union and in the Common Commercial Policy. The question often arose, whether it would have been useful to take some preliminary trade policy measures in that period which lasted until the full integration into the European Union. Analyses have been made of the possible trade and trade policy impact of the customs union and of the Common Commercial Policy, especially in relation to third countries, mainly WTO partners. Sometimes a growing pressure could be felt on behalf of these latter countries. This section elaborates on the dilemmas arising from this connection.

Hungarian MFN tariffs are traditionally higher than the MFN tariffs of the EU. During the previous decade, the difference stabilized at a rate of about two to one. Thus, the average Hungarian level was twice as high. These differences remained at the time of accession. The trade-weighted average of the Hungarian MFN tariff for industrial products at the time of joining the union is estimated to have been 6.9 per cent. At the same time, the similar indicator for the EU is 3.6 per cent. Immediate application of the EU's tariffs from the first day of membership will decrease the level of the Hungarian tariff protection to about half. This obviously improves market access conditions for products originating from MFN partner countries. Their upgraded competitive position will result in higher competition in the Hungarian internal market. It should be mentioned that, at the same time, the relative competitive advantage of products coming to Hungary from the free trade area (first of all, from the EU and CEFTA) will diminish or be partly eroded. In certain cases, this may result in the reorientation of imports towards third countries outside the free trade area.

What has been mentioned above concerns the average level of MFN tariff for industrial products. Obviously, tariffs were dispersed around the average. According to the calculations of some Hungarian experts, about 5,600 Hungarian tariff rates were higher, about 1,600 were lower than the rates of the Common Customs Tariff, and about 350 were identical. Thus, entering the customs union, in some cases, increased the Hungarian rate while in a narrow range of imports there were no changes.

In the group of products where there were EU's tariff rates lower than the Hungarian rate, there were three dominating product groups. The decrease of the level of tariff protection was concentrated on these products. These are electrical machinery and appliances, machinery and equipment, and transport equipment. In the late 1990s, important changes in the product structure and quality upgrades took place in these productions. In most cases, intra-industry and intra-firm trade in the framework of multinational companies became a decisive feature. Consequently, on the one hand, diversion of trade is not very likely. On the other hand, if these large manufacturing companies have and will have imports from MFN countries, these imports become cheaper, thus their competitiveness would improve.

In the group of products where the taking over of the Common Customs Tariff resulted in growing rates, we find mainly raw and basic materials and intermediates. These products are traditionally imported to Hungary from certain Eastern European countries (mainly the Ukraine and Russia), and in many cases with zero duties. According to calculations in this category of products, the average tariff rate is expected to increase from 1.2 per cent to 3.9 per cent after the accession. A disadvantage, on the level of costs, will arise from the tariff increase for Hungarian producers using these products. This disadvantage can be partially compensated if domestic producers are able to transfer these costs to their domestic buyers. This will be facilitated by the weaker import competition stemming from higher external tariffs. Market positions will deteriorate in the case of those producers who utilize such raw materials and intermediates (mainly not alloyed aluminium, chemical materials, light oil for undergoing chemical transformation, methanol, and so on) in products designated for further export. Adoption of the EU's Common Customs Tariffs in these cases may deteriorate the competitiveness of certain processing industries and it would also have an inflationary impact. In order to eliminate or minimize these negative effects during the process of accession negotiations, Hungary formulated a request for temporary tariff quotas for certain products of this category with a lower rate.

Generally, raw materials belong to the category of products where the EU's and Hungarian tariff rates are identical. In many cases this rate is zero in both states. In this sector, there are no consequences of joining the customs union from the point of view of trade patterns.

When speaking of the potential effects of adopting the Common Customs Tariff, one has to take into consideration that, although it will considerably affect some productions, its overall impact on global Hungarian trade will not be decisive. Due to the association agreement with the EU, free trade agreements with other countries and the tariff concessions granted to the least developed countries, less than 20 per cent of Hungarian imports originated in countries subject to MFN tariffs at the assumed time of the accession. Thus, changes due to the shift towards new customs tariffs will affect only a limited part of Hungarian trade. This fact will greatly facilitate the transition to the application of the EU's external tariffs.

Yet, in connection with this transition a predicament may arise; namely, whether Hungary should have begun or accomplished the adoption of the EU's tariffs before the accession. Hungary experienced pressures from both EU and WTO countries to make these steps, either partly or entirely before the date of the accession. In my judgement, these steps should not have been considered.

The main reason is that in the category where such an early transition would have resulted in a decrease in tariff rates (this is the bulk of products), Hungary would have automatically granted unilateral concessions to its MFN partners. It is not obvious or certain that they would have been ready to reciprocate this concession. In this case, there would only have been an increase in Hungarian tariffs at the date of accession. It is therefore better to avoid situations, which are difficult to explain from both an economic and trade policy point of view. Besides this, such a step would have worsened the EU's positions in the enlargement negotiations, especially on the issues of compensation to be held according to paragraph XXIV:C of the GATT 1994 agreement. The EU would not be able to use the argument of tariff rate reduction as a measure of compensating its WTO partners. This does not exclude the possibility that, as a result of the bilateral negotiations, Hungary would have made some tariff reductions to EU level or to zero before the accession. However, this should have been done on a reciprocal basis. (Hungary has entered in such negotiations with the USA.)

If Hungary were to adopt before the accession those external EU's rates that are higher than the original Hungarian level, the position of the affected industries would have deteriorated. This would happen without any compensation on the part of the EU. This would also not have been economically expedient and would be counterproductive from the point of view of those who seek to justify Hungary's request for temporary tariff quotas (Jánszky and Meisel 1999).

Joining the EU also means that Hungary's trade with the member states has become internal trade. Thus, those companies now established in the customs free zones using community imports in order to export to the EU have lost interest in this opportunity. Such companies cover more than 70 per cent of customs free zone enterprises. For them, the zone becomes meaningless.

The remaining 30 per cent of companies are either importing from third countries and exporting to the EU, or importing from and exporting to third countries. Certainly, there are also firms with mixed imports (EU and third country origin) and exports (EU and third country destination). If they want to maintain their status, they have to apply for authorization from the EU. It is difficult to predict the outcome of such requests.

The most likely opportunity to obtain this new authorization lies with those firms that use third country origin inputs for exports towards third countries. In 1999, only four to five companies were carrying out activities of this kind. We may therefore assume that such firms will be able to stay in the customs free zone. In the category of companies using 'mixed' imports for exporting to 'mixed' markets, there are some important and dominant firms in the electronics industry. For them the change of regulations has not been advantageous and there is a possibility that they will leave Hungary.

Even greater difficulties may arise concerning companies that cease their activity in a customs free zone and whether investment goods should be 'internalized', (i.e. customs duties and VAT must be paid). This results from the adoption of the EU's legislation, where import of such goods in the zone is subject to duties. After the accession, the Hungarian government will lose its autonomy of action to grant any exemption.

Changes of these regulatory elements may pose the most serious problems for the lion's share of the companies operating in customs free zones. They may not only lose trust in the Hungarian government, but will also have to face financial difficulties.

'Internalizing' these companies, by removing them from customs free zones and transferring them to the normal customs territory before the accession might be a solution. However, this is not an easy in practice. In this case, the government would be able to decide autonomously whether compensation in the form of an exemption or discharge is necessary or desirable. This solution necessitates quick decisions, information of the firms concerned, and cooperation. Even in this case, technical difficulties connected with taking stock of investment goods and establishing their origin would arise.

In the 1990s, Hungary used safeguard measures as a tool for market protection relatively frequently. This is mainly due to the fact that this tool was the least complicated to introduce. In practice, it was used against imports from countries that were not members of the WTO. It is a fact that in the early 2000s such cases have become less frequent. Nevertheless, this instrument can be considered the unique market protection instrument used during the 1990s. Theoretically, Hungary will retain this possibility after the accession. But there will be differences. Safeguard measures can be applied when market disturbances are proved. The 'threshold' for the launching of these measures will be higher. For example, the magnitude of market disturbance is less on the level of the market of the whole Community than on the small Hungarian market. It will be more difficult to prove such a disturbance for the whole EU, even in the case when traditional Eastern European imports cause difficulties in Hungary. One possible solution would be if Hungary were considered as a region. In this case market disturbances would be assessed on the level of this region. Therefore, Hungary would be in a position to use safeguards in cases when disturbances emerge in only its domestic market. There are precedents to such a solution in the history of the EU.

By taking over the anti-dumping regulation of the EU, Hungary would be able to use this massive market protection measure. As mentioned earlier, despite the existing legislation, Hungary never used anti-dumping measures. By contrast, in the hands of the EU, these are the most frequently applied measure. After the accession, Hungary will have the right and obligation to take over all existing anti-dumping decisions. Thus, the level of protection of the Hungarian market will increase. This affects Hungarian producers differently. Those who produce products that are subject to anti-dumping measures may benefit from the increased protection. For the importers of such products, due to the anti-dumping duties, imports will be more expensive. This may reallocate their imports, most probably towards community suppliers.

## References

Csaba, Laszlo (1996) 'Hungary's Trade Policy and Trade Régime: from Neoprotectionism to Liberalisation or Vice Versa?', Discussion Paper no. 39, July, Budapest: Kopint-Datorg.

Erdős, A.B. and L. Molnár (1999) 'Az árfolyampolitika kérdőjelei' [Question about exchange rate policy], *Cégvezetés*, July.

Inotai, A. (2001) 'Completing Transition: The Main Challenges. The Case of Hungary', in W. Gerstenberger (ed.), *Aussenhandel, Wachstum und Produktivitat – Fragen in Vorfeld der EU – Erweiterung*, Dresden: Ifo Institut.

Jánszky, Á. and S. Meisel (1999) 'Magyarország EU – csatlakozásának WTO – összefüggései' [WTO aspects of Hungary's accession to the European Union], *Európai Tükör Műhelytanulmányok*, no. 64.

Kaminski, B. (1999) 'Hungary – Foreign Trade Issues in the Context of Accession to the EU', World Bank Technical Paper, no. 441.

Kawecka-Wyrzikowska, E. (2000) 'Lessons from Trade Liberalisation in Transition Economies: Ten Years' Experience and Future Challenges', Discussion Paper no. 78, Warsaw: Foreign Trade Research Institute.

Messerlin, P.A. (1998) 'Trade and Trade Policies in Europe, WIIW 25. Anniversary Conference'; Vienna, November 13–15.

Szapári, Gy. and Z.M. Jakab (1998) 'Exchange Rate Policies in Transition Economies: The Case of Hungary', *Journal of Comparative Economics*, December.

Winters, A. (1993) 'A társulási szerződések: egy kis segítség barátainktól [The Association Agreements: With a Little Help from our Friends]', *Külgazdaság*, no. 7.

# 7
# Trade Policy in the Czech Republic in the Decade of Transition

*Miroslav Hrnčíř*

## Introduction

Czech transition from the very beginning has encompassed a simultaneous liberalization in both the domestic and external sectors.

A sweeping liberalization of foreign trade and foreign exchange associated with the introduction of 'internal' currency convertibility for registered businesses on current account transactions was implemented at the very start of the transition, at the beginning of 1991. This came with a package of measures aimed at macroeconomic stabilization and liberalization of domestic prices.

Re-integration into the European and world economy has become a priority for the authorities. The radical opening of the economy was considered a precondition for the success of the entire transition. An early move towards the liberalization of trade and foreign exchange was expected to:

- provide foreign competition discipline and a countervailing power to the monopoly practices of domestic producers and traders
- contribute to a more rational price structure through 'importing' incentives from the world market and through the adjustment of relative prices
- initiate the reallocation of resources reflecting the terms of an open economy.

## The international framework of trade policies

The former Czechoslovakia was one of the founding members of GATT and participated in all the rounds of negotiations that took place in the postwar period. Its framework of a centrally planned economy was, however, inconsistent with the market logic of GATT's rules. As a reaction to its trading system, Czechoslovakia faced considerable trade barriers and less favourable treatment in world markets. Apart from higher tariffs, Czechoslovak exports were often subjected to quotas and other non-tariff barriers, particularly for 'sensitive' items like textiles, rolled material and steel. The records also reveal a substantial number of anti-dumping actions initiated against Czechoslovak exporters.

After the transition to a market type economy began in 1990, the external environment for firms from the former Czechoslovakia also began to change. In April 1990, a new trade agreement based on the MFN principle was signed with the United States. In May of that year, an agreement was made with the EU on trade and economic cooperation, granting higher steel and textile quotas. Since January 1991, until the accession Czechoslovakia enjoyed GSP treatment in the EU, with the exception of 'sensitive' product exports.

### Relations with the EU

By far the most important arrangement was the 'European type' Association Agreement signed with the EU in December 1991. A similar agreement was concluded with the EFTA at the beginning of 1992.

With regard to trade, the main goal envisaged in the Association Agreement was the establishment of a free trade area. The Agreement, however, extended far beyond free trade, covering investment, the legal framework, competition rules, financial cooperation, as well as human and cultural dimensions. Nevertheless, of the four 'basic freedoms', the Association Agreement only provided for the gradual establishment of free movement of goods. There was virtually no commitment to the free movement of labour. After a decade, the issue of free movement of labour from candidate countries remained unresolved. In 2001, with the potential date of entry for the first candidates approaching, the EU was under increasing pressure to find a feasible solution to avoid making that issue a stumbling block for the entire process of accession.

The agreed schedule for the establishment of a free trade area in the Association Agreement was (in principle) based on the non-reciprocity principle. This implies a 'positive asymmetry' in the implementation of tariff reductions and in the elimination of quantitative and other non-tariff restrictions. Other factors conditioning a competitive dominance of EU firms in the mutual trade, particularly in the higher value-added manufactures, were not tackled.

For the 'sensitive' products, like textiles, steel, coal and agricultural products, the establishment of free trade was envisaged by means of special arrangements and protocols (excluding agricultural products, where a mutual concessions approach was to be applied), implying only gradual elimination of tariffs and quotas. Those sensitive products were the dominant Czech exports to EU countries. This implies that the Czech Republic had some comparative advantages under the given circumstances, at least in the short and medium run.

In the course of the 1990s, the full liberalization of trade between the EU and the Czech Republic in industrial goods was, in principle, accomplished (tariffs and quotas). Since January 1996, there have been no tariffs imposed by the EU on Czech exports, while on the Czech side, the remaining tariffs were dismantled as of January 2001. At the same time, technical regulations have been mostly harmonized with EU standards. This provided for access to EU markets free of non-tariff barriers during the pre-accession phase.

**Membership in international institutions and in regional trade arrangements**

The former Czechoslovakia was one of the founding members of GATT. After the dissolution of the Czechoslovak Republic at the end of 1992, the newly formed Czech Republic acceded to GATT in 1993 with no lapses in its application of the General Agreement. The Czech Republic ratified the Uruguay Round Agreements in 1994 and became a member of the WTO. All WTO members and GATT contracting parties are granted at least MFN treatment. The same applies to several other countries with which bilateral agreements containing MFN treatment have been concluded.

The Czech Republic is also a contracting party to a number of regional free trade agreements, which encompass all major trading partners. These include, apart from the EU Association Agreement, the Central European Free Trade Agreement (CEFTA, composed of the Czech Republic, Poland, Hungary, Slovakia, Slovenia, Bulgaria and Romania), and the European Free Trade Area (EFTA).

The Czech-Slovak Customs Union was established on 1 January, 1993, upon the dissolution of the CSFR. The customs union implied common trade and tariff policies in the member countries. Therefore, they shared a common external tariff, including preferential autonomous and bilateral rates. The agreement covered all goods originating in the territories of the Czech and Slovak Republics. Within the customs union, they were not liable to customs duties. On the other hand, goods from third countries did not circulate freely between them. Upon termination of the monetary union between the Czech and Slovak Republics in February 1993, a bilateral clearing arrangement was introduced for commercial transactions between the countries. This arrangement was discontinued in September 1995, when the Czech koruna became fully convertible for current account transactions.

## The trade policy regime in the course of transition

Since the start of the transition, the Czech Republic has maintained a liberal, transparent and relatively stable trade policy.

The trade regime is based largely on customs tariffs and quotas, with few non-tariff barriers in use. Customs tariffs apply only to imports, and no export tariffs are levied. Apart from these, indirect taxes are levied on imports (value added tax and excise taxes on selected goods, the rates of which are the same for domestic and imported products in conformity with international commitments). All tariffs are currently applied at their bound rates and no specific, composite or other non-ad valorem tariffs are used, making the external tariff transparent.

Prior to the start of the transition process, quantitative restrictions were the instrument for controlling trade. Tariffs were collected on trade within the convertible currency area, but did not play a significant role as a policy instrument. The inherited tariff structure reflected this. Therefore, in the first years of the transition, the decision was made to restructure the customs tariff schedule. The aim was to align the tariff level with the EC and neighbouring countries and to correct for the distorted tariff structure by reducing the level of trade protection for raw or semi-processed materials, components and spare parts. The corresponding GATT procedures and negotiations were invoked and successfully completed in December 1992.

At the beginning of the transition, the simple average tariff was 5.9 per cent and the import-weighted average rate was 5 per cent. Most tariff positions were bound in GATT, including all tariffs on agricultural products. The trade protection offered by the previous tariff schedule was very low, not only in absolute terms, but also in comparison with other countries. The tariffs were below the EC level and also below the average tariff levels of Poland and Hungary. The data reflecting the developments in the course of the past decade show that these properties of the tariff system were retained throughout the period. After the restructuring in 1992, the average tariff level increased only marginally (less than 1 per cent) and participation in the Uruguay Round resulted in further tariff reductions. The average bounded tariff incidence amounted to 4.5 per cent in the year 2000, 3.6 per cent for industrial goods and 10 per cent for agricultural and foodstuff products.

The Czech Republic's MFN tariffs are not only low; they are also fairly uniform across the various economic sectors. Exceptions exist in agriculture which, following the Uruguay Round tariffication, still had a few items with high tariffs. Food products and beverages have tariffs averaging nearly triple the level elsewhere in the economy. In general, the MFN tariffs reveal some escalatory trend – the higher the level of processing, the higher the average tariff level. This is particularly noticeable in sectors such as food processing, textiles and clothing.

In the first half of the 1990s, the Czech Republic had no anti-dumping laws, countervailing duty, or safeguard legislation in force. Only in 1997 was a law concerning protection against importation of dumping products passed. Further, as of 1 July, 2000, two important pieces of legislation have entered into force. The first piece was a law on various measures regarding the import and export of goods and licensing procedures. The second was concerned with protection against the import of subsidized goods and the violation of WTO rules. Both laws enforced the commitments with respect to the WTO in the various areas.

In the course of this chapter, cases of backsliding from the existing liberalization level in trade and capital flows are examined, as well as arguments which were raised to justify the reversals or the temptation to resort to them. Of course, the concrete situation and causal factors diverged in the course of time. The Czech economy went through a number of stages in the 1990s, in which both external and domestic conditions underwent substantial change and development. Accordingly, it seems useful to distinguish between three periods:

- 1990–1992: The initial transition stage: sweeping price, trade, and foreign exchange liberalization
- 1993–1997: Robust growth, currency crisis, and follow-up recession
- 1998–2000: From recession to accelerating growth, from a current account balance to surging deficits?

## The macroeconomic setting and trade policies in the initial transition stage of 1990–92

### The macroeconomic setting in the early transition years

The political, institutional and systemic changes initiated by the Velvet Revolution of November 1989 created an entirely new environment for economic activities, including trade, capital and foreign exchange flows. Unleashed by the dissolution of prior state controls on the foreign trade and foreign exchange systems, and by the sweeping domestic price liberalization (associated with liberalization in foreign trade and foreign exchange spheres), a radical adjustment in trade, services and capital flows began to materialize. This was accompanied by the introduction of limited 'internal' currency convertibility at the beginning of 1991. This process initiated the shift from the distorted structures, which arose under the centrally planned economic framework, to a market determined pattern of trade and capital flows, reflecting underlying comparative advantages.

However, the years 1990–92 could simply represent the first stage of adjustment. The policies in this period were dominated by macroeconomic stabilization and liberalization, both in domestic and external spheres. The privatization programme and the process of enterprise restructuring were only at their inception. The implementation of the systemic and legal changes, related especially to the progress of privatization, could be only initiated. Reflecting this 'macroeconomic' character of the first transition years, most changes that materialized in trade and capital flows during this period resulted from the macroeconomic shifts and macroeconomic policies. These, in turn, were due to the impact of various exogenous shocks. The microeconomic causes of trade and capital flow development implied in the changing behavioural pattern of economic agents became more pronounced only in the later stages.

### Developments in trade flows

In the period between 1989 and 1992, the relative export performance of the Czech Republic, as reflected in its share of world exports, continued to

*Table 7.1*  Czech Republic: Czech Republic's share in world exports

| Year | World exports (USD mn) | Czech exports (USD mn) | Czech share of world exports (per cent) |
|------|------------------------|------------------------|-----------------------------------------|
| 1989 | 30206 | 12350 | 0.41 |
| 1990 | 35470 | 8317 | 0.23 |
| 1991 | 36122 | 7979 | 0.22 |
| 1992 | 37226 | 8227 | 0.23 |

*Source*:  Statistical survey of exports and imports of the Czech Republic, *International Financial Statistics*.

decrease. As follows from Table 7.1, the export volume of the Czech Republic in terms of USD had dropped by a third since 1989, thus accelerating the long-term decline of ex-Czechoslovakia's share of world markets.

Yet the identified trends in trade volume must be interpreted in connection with the underlying changes and interruptions, both domestic and external, which were specific to the period between 1990 and 1992. The impact of a severe contraction of domestic demand and of an unexpectedly deep fall in the overall economic activity in the aforementioned years interacted with the impact of various external shocks. This resulted in the collapse of CMEA trade and institutions, and in the beginning of a parallel process aimed at the redirection of trade to Western markets. The data proves that the extent of the shift of trade to developed market economies during the period was remarkable, despite only a gradual phasing out of various trade barriers on the side of western trade partners.

Parallel to the contracting trade volume, the share of higher value-added products in total Czech exports continued to diminish between 1989 and 1992. Given that SITC groups 0–5 represent raw materials and SITC groups 6–8 finished products, the share of raw materials soared from roughly one third to two thirds of total exports, while the share of finished products fell accordingly (cf. Table 7.2). In particular, a dramatic decrease materialized in the share of machinery and transport equipment (SITC 7), with exports dropping from 47 per cent in 1989 to a mere 25 per cent in 1992.

However, these developments must be assessed and interpreted in the context of a massive trade redirection. Due to the 'dual' structure of Czechoslovak trade prior to the 1990s, the reallocation of exports to developed market economies could be expected to result in, at least for a period, an increased share of lower value-added products in its total exports. At the same time, an encouraging trend in the pattern of exports to the developed market economies has started to evolve. The role of finished products (SITC 6–8) had been steadily increasing since 1990. Of particular importance were the gains in the export share of machinery and transport equipment (SITC 7), especially to EU economies.

*Table 7.2* Czech Republic: Czech export and import patterns, 1989–92 (percentage shares)

|  | **Materials** | **Finished products** |
|---|---|---|
| Export |  |  |
| 1989 | 38.7 | 61.3 |
| 1990 | 37.9 | 62.1 |
| 1991 | 58.2 | 41.8 |
| 1992 | 63.0 | 37.0 |
| Import |  |  |
| 1989 | 53.4 | 46.6 |
| 1990 | 52.6 | 47.4 |
| 1991 | 60.0 | 40.0 |
| 1992 | 52.6 | 47.4 |

*Source*:   *Facts on Czechoslovak Foreign Trade*, Ministry of Trade and Industry.

*Notes*:   Materials = SITC 0–5
Finished products = SITC 6–8

## Trade and payment balances in 1990–1992

As shown in Table 7.3, after a sharp deterioration in 1990, the current account balance in convertible currencies recovered in 1991 and 1992, and recorded a surplus in both years. Compared to the past, when merchandise flows were dominant, a tendency towards an increased share of services,

*Table 7.3* Czech Republic: balance of payments in convertible currencies, 1990–92 (in millions USD)

|  | **1990** | **1991** | **1992** | |
|---|---|---|---|---|
|  | **CSFR** | **CSFR** | **CSFR** | **CR** |
| Current account | −1105 | 356.5 | 225.6 | 52.9 |
| Trade balance | −785 | −447.4 | −1575.6 | −1372.2 |
| Service balance | 37 | 827.4 | 1652.4 | 1314.2 |
| Income balance | −316 | −65.4 | 8.8 | 5.9 |
| Transfers | −40 | 41.8 | 140.0 | 105.0 |
| Financial account | 326 | 47.4 | 40.6 | −6.1 |
| FDI | 181 | 592.4 | 1054.9 | 982.9 |
| Portfolio investment | − | − | −42.6 | −35.8 |
| Other long term capital | 718 | 1731.7 | 471.4 | 320.5 |
| Short term capital | −573 | −2277.1 | −1443.1 | −1273.7 |
| Errors and omissions | −324 | 494.4 | −386.3 | −126.9 |
| Change in reserves ( - increase) | 1102 | −897.9 | 120.1 | 80.1 |

*Source*:   Czech National Bank.

tourist trade, and income transfers was clearly surfacing and contributing to favourable results on the current account balance.

## The 'infancy' stage of capital flows

Given the underdeveloped financial infrastructure, the lack of full foreign exchange and the financial account liberalization, capital flows were mostly confined to long-term official capital inflows combined with the occasional short-term capital flight.

Apart from the first wave of foreign direct investment related to privatization, the long-term inflows were the result of the endeavours of monetary authorities to replenish the disappointingly low levels of foreign exchange reserves. At the same time, they reflected the support extended by international institutions to the initiated process of transition.

In contrast to official capital inflows, private investors reacted through short-term capital outflows to the conditions of uncertainty, shocks and devaluation expectations of the initial transition stage. The country's international reserves were put under pressure, especially when the waves of capital flight emerged. The first flight occurred in the second half of 1990, prior to the implementation of the sweeping liberalization package linked to the 'pre-announced' devaluation of the Czechoslovak koruna. The second flight developed two years later, once the forthcoming split of the federal republic and the common currency became evident. In both cases, capital flight materialized, in spite of the relatively rigid capital controls.

These short-term capital outflows were not equally matched by long-term inflows and, as a result, the cumulative financial account in convertible currencies showed a deficit for the period 1990–92.

## The exchange rate regime

The types of policies suggested for adoption at the beginning of the transition process diverged widely. This was a result of the priority and weight attributed to the competing objectives of competitiveness and macroeconomic stability. Reacting to the uncertainties in the assessment of the current situation with regard to the country's competitiveness and future developments, the flexible option of the exchange rate was put forward. When the emphasis was attached to restoring macroeconomic stability, the pegged exchange rate option was claimed to be the right choice.

Powerful arguments for exchange rate flexibility related to the conditions of transition were as follows:

- The inherent features of transition were the discontinuity with the past, and uncertainty as to what a reasonably competitive exchange rate level could be.
- The transition process itself, apart from exogenous shocks, implied turbulent developments due to extensive liberalization, adjustment, and

restructuring in the domestic economy. In these conditions, it seemed inevitable that the persistent disparity with fundamentals in partner countries would last for a protracted period, in particular, a higher level of inflation.

- Apart from these disparities, the pre-requisites for a sustainable fixed exchange rate regime seemed to be lacking, in particular, a sufficient level of international reserves and available funds to defend a fixed exchange rate, a workable infrastructure of financial markets and credible monetary policies.

The floating rate was suggested as a way to circumvent uncertainty about the level securing competitiveness and to avoid the risk of foreign exchange constraint through an unduly undervalued or overvalued rate. It should have also allowed for macroeconomic policies to be calibrated without worrying about the competitive level of the exchange rate.

Nevertheless, the arguments in favour of a fixed exchange rate regime were equally compelling:

- A fixed exchange rate regime and nominal exchange rate stability were considered to be the proper instruments of macroeconomic stabilization in an economy that was implementing a sweeping liberalization. With this mode of reasoning, the exchange rate was expected to take over the role of a key nominal anchor.
- A pegged exchange rate could provide a benchmark for an extensive price ratio adjustment triggered by price and foreign exchange liberalization.
- A fixed exchange rate regime could impose a type of stabilizing impact on the economic agents, firms and trade unions.
- The traditional problems associated with a floating rate, especially the high volatility and risk of a substantial misalignment, were likely to be much greater than in a consolidated market economy. This was due to the weak currency markets, underdeveloped institutions lacking foreign exchange markets, and short-term arbitrage flows, which could have worked towards stabilizing the exchange rate. The instability of a key price or of the exchange rate could have undermined the main goal of currency convertibility. This goal was to contribute to the adjustment of price ratios via their 'importing' and to the disciplining of domestic agents.

The decision of the authorities was to opt for a pegged exchange rate regime. Its adoption was substantially influenced by the IMF's stabilization programme for countries in transition, which included Czechoslovakia. At the time, they were advocates of the nominal anchor approach. Within this programme, the stability of the fixed exchange rate was targeted to be a major instrument of macroeconomic stabilization and to anchor both current flows and expectations.

*The exchange rate level*

Given the pegged exchange rate solution, the monetary authorities were confronted with the selection of a proper 'entry' exchange level rate. The issue had two dimensions:

- What its equilibrium level might be
- How far to deviate from such a level to secure exchange rate stability over a certain 'desirable' period, given the persistent but expected disparity in the fundamentals.

Though a reliable assessment of the equilibrium rate was evidently not feasible, there were calls for an initial exchange rate undervaluation. An example is the deliberate deviation of the actual rate from the equilibrium rate. What the currency convertibility aimed for would be otherwise hard to sustain with a pegged exchange rate regime. The likely outcome would be a rapid realignment of the exchange rate, with a possible retreat from the established convertibility status.

Additional constraints made the requirement of entry undervaluation even more appealing:

- The level of foreign exchange reserves being too low to provide a desirable cushion for potential fluctuation in liberalized foreign exchange flows
- The need to reorient the bulk of existing trade after the collapse of Comecon institutions and markets
- Persistent, real exchange appreciation, which was expected from an economy in transition.

The main bone of contention, though, turned out to be the extent of depreciation with which to start. There was, however, an obvious trade-off. A larger undervaluation would increase the chances of survival of the introduced exchange rate. On the other hand, it would deepen inflationary cost-push pressures and mitigate the disciplining function of the exchange rate. Underestimated depreciation would result in unsustainable trade and current account deficits and possibly protective measures, undermining the credibility and the very existence of currency convertibility. The latter case implied the failure of the liberalization and reform programme, which would have been considered more detrimental. The resolution of the authorities was if an error is to be made, then it would be better to have overshot undervaluation than the other way round.

In assessing the extent of depreciation, the options on the pace of liberalization in the external sphere were of utmost importance. The different views suggested alternative approaches for:

- the advancement of foreign trade and foreign exchange liberalization
- the level and time profile of changes in trade protection
- the timing of the introduction of currency convertibility.

According to one approach, priority ought to be given to domestic liberalization, while the regulations on the external sphere should be dismantled in due time. The alternative approach, which was eventually implemented, argued for a sweeping liberalization on the domestic and external fronts simultaneously. Obviously, the advocated depreciation for the latter case was higher.

The degree of uncertainty and the lack of consensus as to the proper entry level for the exchange rate were reflected in a wide range of values suggested for its 'initial' level. These values ranged from CSK 16 per USD (i.e. close to the then existing 'commercial rate') up to CSK 35–38 (i.e. the rates approaching the black market rates). The option implemented, CSK 28 per USD, was closer to these shadow rates. The implied depreciation vis-à-vis the previous commercial 'coefficient' amounted to 95 per cent. The CSK exchange value thus decreased to 56 per cent of its former 'commercial' level.

The effected depreciation widened the pre-existing wedge between the actual exchange rate and the purchasing power parity rate. Such a gap between the domestic and external values of their currencies was a common feature of former centrally planned economies. This occurred in varying degree amongst all emerging market economies. Given their relatively low level of development, they faced productivity and efficiency gaps with respect to their developed partners. Nevertheless, the extent of CSK entry depreciation (as well as that of the Polish zloty) was criticized as substantially unrealistic. The fact that it was possible to maintain the nominal exchange rate of the Czech koruna (CZK) at the same level from January 1991 up to the currency crisis of May 1997, despite persistent real appreciation, was claimed to be a sign of initial excessive undervaluation.

However, the assessment of the costs and benefits of entry devaluation must take into account the specific conditions of transition, and discriminate between short- and medium-run effects. Maintaining and further widening the disparity between domestic and external values of CSK was a viable way of securing export competitiveness in the short run. This was important in a period when export growth to convertible currency markets was considered crucial to offset the sharp contraction of trade to the former Comecon countries, in the hope of maintaining the export revenue used to finance expanding imports of western technology and other products. At the same time, however, an exchange rate 'cushion' diminished the hopes for the disciplinary impact of the fixed exchange rate. Moreover, the performance environment of domestic firms with respect to the outside became 'softer'. The ensuing cost advantages provided a basis for export of those items in which the costs and price dimensions of competitiveness dominate. These are mostly standard, lower value-added products. The implied incentives were thus likely to support the shift to 'cheap labour' advantages and to low value-added production and export patterns. This type of setting, if maintained, was likely to work against the desirable properties of competitiveness in the medium and long run.

The extent of depreciation must have contributed to a higher than expected price level jump in the first months of 1991. Some 35 per cent–40 per cent of the surging price level in that period was estimated to be the result of currency depreciation.

## Trade policy measures to cope with implied uncertainty

In the above-described macroeconomic setting and while liberalizing the trade system substantially, some trade policy measures were adopted by the authorities of the former Czechoslovakia to guard against a potential surge in imports. The reason for this was to guard against an unsustainable trade balance and current account development.

The initial stage of the transition process, from 1990 to 1992, was dominated by substantial interruptions, external shocks, and domestic restructuring. Given the implied uncertainties and the forecasts of future macroeconomic developments, the trade and payment flows were subject to considerable margins of error. For example, government forecasts envisaged for 1991 a fall in GDP in the range of 5 per cent–6 per cent and a current account deficit around USD 2.5 bn. In reality, the decline in GDP was much greater while the current account registered a small surplus of USD 0.3 bn instead of a large deficit.

In the trade sphere, the information problem was further accentuated by institutional changes. Every citizen was now entitled to engage in international business, and so the number of new entrants in foreign trade mushroomed. At the same time, systemic changes (e.g. the shift to customs statistics in the beginning of 1991) complicated the consistency of time series data. As a result, the partners' statistics (e.g. OECD data) diverged substantially from Czechoslovak sources.

There were uncertainties on how imports and exports would react to the changing macroeconomic environment, liberalization and to a new market-based trade regime. The main arguments for the introduction of protectionist measures were of a precautionary character. The corrective measures were targeted to avoid the emergence of substantial payment imbalances. This would have undermined the introduced policy of a fixed exchange rate as a nominal anchor. They were also expected to provide domestic producers with some time span for adjustment after the abrupt opening of the economy. The forms of protective measures adopted were as follows:

### An import surcharge

This was introduced in December 1990. In its notification to GATT, the Czech authorities declared that it should be considered as a temporary measure. This was to be re-examined when the outcome of the trade balance became clear, at the end of 1991.

Initially, the import surcharge of 20 per cent, calculated on the basis of custom values, was applied mainly to consumer goods. In April 1991, some provisions concerning the implementation of an import surcharge were modified (exemptions granted for goods imported for personal use and for goods imported for production). In May 1991, the rate of the import surcharge was reduced to 18 per cent, and then to 15 per cent in June 1991. By mid-1992, it was lowered to 10 per cent. Decreased step by step, the surcharge was completely phased out by the end of 1992, and before the dissolution of Czechoslovakia.

### A licensing system

The old system of quotas and licensing, created under central planning, was transformed into a transparent system of licensing. This applied to certain categories of trade with explicit prohibitions and quotas during 1990 and 1991. On the import side, the licensing concerned four categories (crude oil, natural gas, narcotics, and arms and ammunition). On the export side, the system covered three main categories: sensitive goods (arms, weapons and explosives); items the import of which was subject to quantitative restrictions by partner countries; and some 125 food and intermediate products for which there was concern about the risk of shortages and excessive price increases on the domestic market.

The first two categories of export licensing were similar to practices utilized by other countries. The third category, food and intermediate products, remained under some price regulation which was a transitional measure phased out after undergoing further price liberalization steps.

### A global import quota for selected agricultural products

A global import quota for selected agricultural products was introduced in September 1991, backed by the reasoning that their actual very low world prices were causing serious injury to domestic producers. This restriction had been eliminated by the end of the year but was replaced by restructured customs tariffs.

## Robust growth, the currency crisis and follow-up recession in 1993–97

### Macroeconomic developments and policy reaction

During this period, the Czech economy recovered from the crisis of the initial transition years and went through a period of relatively robust growth. However, this ended with a currency crisis and a subsequent recession (cf. Table 7.4). The underlying causes of this outcome were domestic imbalances. The surging domestic aggregate demand was only partly met by the domestic supply side response. The data shows that the increasing excess of demand over domestic supply was channelled to trade and current account

*Table 7.4*   Czech Republic: GDP growth and components of aggregate demand, 1993–97 (y/y change in %, constant 1995 prices)

|                        | 1993  | 1994  | 1995  | 1996  | 1997  |
|------------------------|-------|-------|-------|-------|-------|
| GDP                    | −3.2  | 4.8   | 5.9   | 4.3   | −0.8  |
| Household consumption  | 3.0   | 8.0   | 5.8   | 7.9   | 2.4   |
| Public consumption     | 8.3   | 0.4   | −4.3  | 3.6   | −4.4  |
| Gross fixed investment | 32.9  | 19.3  | 19.8  | 8.2   | −2.9  |
| Exports of goods       | 16.6  | 49.5  | 17.0  | 6.2   | 13.7  |
| Services               | −     | −     | 15.7  | 14.2  | −3.5  |
| Imports of goods       | 15.3  | 51.6  | 27.0  | 12.5  | 10.7  |
| Services               | −     | −     | 0.2   | 17.3  | −3.4  |

*Source*:   Czech Statistical Office.

*Table 7.5*   Czech Republic: balance of payments, 1993–97 (in CZK bn)

|                                      | 1993  | 1994  | 1995   | 1996   | 1997   |
|--------------------------------------|-------|-------|--------|--------|--------|
| Current account                      | 13.3  | −22.6 | −36.3  | −115.5 | −101.9 |
| Trade balance                        | −15.3 | 39.8  | −97.6  | −159.5 | −144.0 |
| Service balance                      | 29.5  | 14.1  | 48.9   | 52.2   | 55.9   |
| Income balance                       | −3.4  | −0.6  | −2.8   | −19.6  | −25.1  |
| Transfers                            | 2.6   | 3.6   | 15.2   | 10.4   | 11.3   |
| Financial account                    | 88.2  | 97.0  | 218.3  | 113.6  | 34.3   |
| FDI                                  | 16.4  | 21.6  | 67.0   | 34.6   | 40.4   |
| Portfolio investment                 | 46.7  | 24.6  | 36.1   | 19.7   | 34.4   |
| Other long-term capital              | 23.5  | 31.9  | 89.4   | 84.4   | 12.9   |
| Other short-term capital             | 1.6   | 19.0  | 25.8   | −25.2  | −53.5  |
| Errors and omissions                 | 3.0   | −6.1  | 15.8   | −19.6  | 11.2   |
| Change in reserves ( - increase)     | −88.3 | −68.3 | −198.0 | 22.5   | 56.0   |

*Source*:   Czech National Bank.

deficits rather than to higher inflation (see Table 7.5). On the external front, massive capital inflows exasperated rising imbalances. They enabled current account deficits to persist and deepen, simultaneously contributing to persistent exchange rate appreciation, despite the mentioned trade and current account deficits.

The reaction to the increased overheating of the economy and to external imbalances was deficient. Fiscal policy and wage developments were not adjusted in due time. Capital inflows and excessive wage increases were the main culprits of the evolving situation. The lack of response in fiscal and wage spheres was the single most evident failure of macroeconomic policy setting at the time. Only the monetary policy was tightened in mid-1996. However, given the underlying causes of the imbalances, its impact could only be one-sided. Moreover, monetary tightening added an additional attraction to speculative inflows. These were, to some extent, counterproductive side effects of the policy move in the conditions of an open financial account. Consequently, the evolving policy mix did not meet the requirements of the given macroeconomic situation.

The reluctance to adjust was mostly due to the political cycle. In 1996, there were two rounds of general parliamentary elections, for the lower and upper chambers. In addition to this, there was an apparent underestimation of the requirements of the 'rules of the game' in a small, open economy. As a result, a more comprehensive policy reaction to the unsustainable macroeconomic development came rather late, only under the pressure of an already developing crisis. Consequently, during the crisis, the use of restraint proved to be excessive and contributed to a protracted recession in the period 1998–99.

However, macroeconomic policy slippages were not the only cause of overheating and increasing imbalances. Instead, the acceleration of growth rates proved unsustainable due to the microeconomic, institutional, and legal defects and rigidities. Though mostly inherited from the pre-transition past, they lacked proper cultivation and policy attention. This was apparent in the stalled privatization process, especially in the case of the banking sector.

The main factors of the currency crisis of May 1997 and of the subsequent recession were 'home-made', with both the macroeconomic and institutional factors contributing. In the conditions of massive capital flows and an open financial account, these domestic defects made the Czech economy vulnerable and susceptible to external contagions. The massive speculation against the CZK triggered the crisis, but simply reflected the implied domestic weaknesses.

### Advances in the liberalization of external flows

In the first years of this period, the liberalization of external and foreign exchange flows made further advances. The new foreign exchange act (no. 219/1995) which came into force on 1 October, 1995 provided for the current account convertibility status in accordance with Article VIII of the IMF and, at the same time, went beyond that status by extending liberalization to a number of capital account items.

Prior to the 1995 exchange act, a few liberalization steps were implemented and included:

- easing the transfer requirement of legal persons and the extension of their right to keep foreign exchange accounts with domestic banks (1994)
- liberalization of the sale of domestic securities to non-residents (1994)
- an increase in the limit of foreign exchange purchases by tourists to up to CZK 100 000 (1995)
- liberalization of FDI with EU countries (1995).

The new Foreign Exchange Act consolidated these steps as well as the actual status of liberalization – all well in advance of existing formal regulations. It abolished the remaining restrictions on the current account and any regulations limiting the access of citizens to foreign exchange. The export and import of CZK and unrequited transfers were fully liberalized.

Beyond current account transactions, the act also liberalized some other capital flows. Deregulation progressed more on the side of inflows, covering among other things financial credits from non-residents. Several types of outflows were also liberalized. Among them were:

- FDI by residents
- purchase of real estate abroad by residents
- substantial easing of rules for purchasing of foreign securities by domestic entities.

In the end, the Foreign Exchange Act of 1995 consolidated the implied liberalization status and represented a further step towards a fully opened financial account. The remaining restrictions on capital flows were mostly abolished in the course of the subsequent period. As of the beginning of 2001, this applied to:

- full liberalization of the purchase of foreign securities
- the issue of foreign securities in the Czech Republic and their introduction to the domestic market
- extending credits and guarantees by residents to non-residents
- opening bank accounts by residents abroad.

There was an exception for non-residents in capital participation in a few selected institutions and areas. These were banks, auditing institutions and the stock exchange. The only significant remaining regulation of capital flows applies to the purchase of real estate by non-residents.

### Capital flows

Since the second half of 1993, the importance, volume, and structure of capital flows have substantially changed (see Table 7.6). There was a successful stabilization of the economy and privatization was advancing. The liberalization of both the external flows and the domestic economy offered incentives to investors and to capital inflows. However, there were calls for increased capital inflows due to the constrained domestic supply of financing, in particular of long-term credits. There was also a much wider interest rate differential in comparison with the standard in the developed world. That differential was linked with the nominal stability of the pegged exchange rate, which had already earned some credibility and provided a 'one-way bet' for speculators. The causes of capital inflows were a result of the success of stabilization packages and the implied imbalances and constraints of the transition process. This offered an attractive margin of profit until the imbalances were corrected.

*Table 7.6* Czech Republic: external investment position, 1993–97 (in USD bn, end of period)

|  | 1993 | 1994 | 1995 | 1996 | 1997 |
|---|---|---|---|---|---|
| Assets | 18.0 | 20.4 | 29.4 | 30.6 | 29.8 |
| FDI | 0.2 | 0.3 | 0.3 | 0.5 | 0.5 |
| Portfolio | 0.3 | 0.4 | 0.8 | 1.4 | 1.0 |
| Other long term capital | 8.4 | 8.2 | 8.1 | 8.6 | 8.5 |
| commercial banks | 0.1 | 0.1 | 0.2 | 0.9 | 1.1 |
| government | 6.3 | 6.3 | 6.0 | 5.9 | 5.9 |
| other sectors | 1.2 | 1.0 | 0.9 | 0.8 | 0.7 |
| Other short term capital | 5.3 | 5.3 | 6.2 | 7.7 | 9.9 |
| commercial banks | 2.8 | 2.9 | 3.3 | 4.7 | 7.2 |
| other sectors | 2.3 | 2.4 | 2.9 | 3.0 | 2.7 |
| Liabilities | 14.1 | 18.1 | 27.2 | 33.2 | 32.9 |
| FDI | 3.4 | 4.5 | 7.4 | 8.6 | 9.2 |
| Portfolio | 2.0 | 2.9 | 4.7 | 5.3 | 4.9 |
| Other long term capital | 6.4 | 7.4 | 10.5 | 13.7 | 12.3 |
| commercial banks | 0.5 | 0.9 | 3.4 | 5.2 | 4.1 |
| government | 2.7 | 2.7 | 2.0 | 1.6 | 1.1 |
| other sectors | 2.0 | 3.7 | 5.0 | 6.8 | 7.0 |
| Other short term capital | 2.3 | 3.2 | 4.6 | 5.6 | 6.4 |
| commercial banks | 0.7 | 1.5 | 2.6 | 3.7 | 4.8 |
| other sectors | 1.4 | 1.6 | 2.0 | 1.8 | 1.6 |

*Source*: Czech National Bank.

Unlike in the past, private capital flows started to dominate, and massive inflows resulted in an unprecedented financial account surplus (cf. Table 7.5). The GDP data available from that time shows the ratio of capital inflows to GDP amounting to 9.4 per cent in 1994 and 18.4 per cent in 1995. After the revisions the statistical office made in time series on GDP in 1997 and again in 2001, a modest downward correction of the mentioned ratios resulted. According to those data, net capital inflow amounted to 8.2 per cent in 1994, 15.8 per cent in 1995 and 7.2 per cent in 1996.

*Foreign direct investment*

The entire inflow of FDI to the Czech Republic in the period between 1990 and 1997 amounted to USD 9.2 bn, of which USD 5.2 bn materialized within the years, 1994–96. Compared to some other countries in the region, this inflow wave was registered as a record result. Notably, the Czech authorities, unlike some other transition economies, did not provide any special incentives (e.g. tax holidays) to foreign investors during the given period.

*Short-term capital*

Short-term capital flows increased considerably during this period. These inflows were attracted by a large interest rate differential[1] combined with a

*Table 7.7* Czech Republic: external debt and foreign exchange reserves, 1993–1997 (in USD bn, end of period)

|  | 1993 | 1994 | 1995 | 1996 | 1997 |
|---|---|---|---|---|---|
| External debt (convertible currencies) | 8.5 | 10.7 | 16.5 | 20.8 | 21.4 |
| long-term | 6.5 | 7.8 | 11.5 | 14.8 | 14.3 |
| short-term | 2.0 | 2.9 | 5.0 | 6.0 | 7.1 |
| (inconvertible currencies) | 1.1 | 1.5 | 0.6 | 0.3 | 0.3 |
| Official reserves (CNB) | 3.9 | 6.2 | 14.0 | 12.4 | 9.8 |

*Source*: Czech National Bank.

fixed CZK exchange rate. Until the brink of the currency crisis in May 1997, the expectations assumed a continuation of nominal exchange rate appreciation. As a result, the share of short-term debt increased from 23 per cent in 1993 to almost 34 per cent in 1997 (cf. Table 7.7).

*Other long-term capital*

Capital flows in this category reflected a rapidly increasing volume of credits contracted by Czech entities (companies, banks, municipalities). The incentives behind this surge of flows were manifold. A strong incentive was the limited supply of medium- and long-term financing by domestic banks, while foreign credits became increasingly available for creditworthy (mostly large) clients.[2] Further, in an environment of nominal exchange rate stability, Czech entities mostly disregarded the implied exchange rate risks and, consequently, foreign credits seemed to be cheaper and preferable.

### The controversial impact of capital flows

The massive capital inflows during the period 1994–96 and their consequences represent a controversial phenomenon in the Czech Republic. On the one hand, capital inflows provided a valuable contribution to domestic savings. This enabled an economic recovery and increased the rate of growth. The substantial use of foreign savings seemed justified given the 'inherited debts'. These were the under-capitalization of the economy due to the existence of obsolete technology and an underdeveloped infrastructure. There were also costly requirements for environmental protection and extensive restructuring. It must be noted, however, that the domestic savings ratio to GDP fluctuated well above the international average, amounting to 29.3 per cent in 1995 and 27.8 per cent in 1996 (see Table 7.8).

In contrast, capital inflows also implied risks and increased the vulnerability for the national economy. Apart from their impact on the money supply and the related costs of sterilization, the relief gained from foreign exchange constraint allowed domestic borrowers to increase their indebtedness. Consequently, the potential risk for the national economy kept increasing.

*Table 7.8* Czech Republic: current account savings and investment ratios to GDP, 1995–97

|  | **1995** | **1996** | **1997** |
|---|---|---|---|
| Current account / GDP | −2.6 | −7.4 | −6.1 |
| Domestic savings / GDP | 29.3 | 27.8 | 26.6 |
| Gross capital formation / GDP | 34.0 | 34.2 | 32.6 |
| Fixed capital formation / GDP | 32.0 | 31.9 | 30.6 |

*Source*: Czech Statistical Office.

Due to the growing volume of volatile short-term flows, the share of total debt increased by one third in 1997. This was exacerbated by the vulnerability of the national economy to external shocks, contagions and the changed expectations of investors.

There were two waves of short-term flow reversals in the discussed period. The first occurred in the first half of 1996, after widening the exchange rate fluctuation band. The second and more significant wave occurred during the currency crisis of May 1997. The speculative short-term flows and short-selling triggered the currency crisis. Consequently, net capital inflows, which amounted to almost 16 per cent of the GDP in 1995, shrank to just 2 per cent in 1997.

### Measures to control capital inflows

While coping with the mass capital inflows and their consequences during the mid-1990s, a host of various policy measures were also applied. Some of them were relatively quick and easy to impose. These were suitable for 'the first line of defence', but their potential was soon eroded. On the other hand, policies aimed at an increased absorption capacity of the economy, such as structural policies, required more time in order to become effective. The measures for the 'first line of defence', apart from sterilization, included liberalization of outflows (discussed above). This was accompanied by some other steps to slow down the capital inflows. The aim was to withstand, at least temporarily, the volatile and speculative short-term flows.

In April of 1995, the central bank introduced a fee (margin) of 0.25 per cent on its transactions in foreign exchange with commercial banks. Given the existing wide interest rate differentials, such a 'grain of sand' proved to be of negligible importance.

In August of 1995, a quantitative limit was placed on the net short-term open positions (liabilities) of banks to non-residents by the central bank. These positions were targeted not to exceed either of the two ceilings: 30 per cent of the bank's balance sheet short-term assets and liabilities and/or an amount of CZK 500 million. The rationale behind this measure was to discourage the massive short-term inflows. However, this could be relatively easily circumvented (through reclassifying the maturity of contracts to more than one year or through engaging a domestic non-bank financial entity as an intermediary). As a result, the measure proved to be increasingly ineffective over time.

The objective to secure relatively quick control over the wave of capital inflows was also reflected in the provisions of the Foreign Exchange Act. The introduction of the option for obligatory deposits followed the experience of some other countries. Their eventual imposition was viewed as a potential defence measure in order to tax the capital inflows, short-term ones in particular. Despite the massive inflows of volatile capital in 1995 and 1996 (changing to outflows later on), the Czech authorities never enacted this provision. Later on, the provision was formally suspended, under the pressures of its various OECD commitments.

The foreign exchange act also had clauses for government interference in the case of a critical balance of payment problems or a serious risk to currency stability. In such a case, the government was entitled to suspend the validity of the Act for six months and interfere with foreign exchange markets and capital flows. This emergency arrangement was a standard practice within the international context, comparable to those existing in some other OECD countries.

## Exchange rate developments and the exchange rate regime

The pegged exchange rate regime was maintained, together with the stability of the nominal exchange rate, until May 1997. This was a distinguishing feature of Czech developments in contrast to most economies in transition. This was achieved despite persistent real rate appreciation and external shocks.

A very narrow fluctuation band of ±0.5 per cent to a currency basket was maintained until February 1996. Yet the composition and the relative weights of the basket currencies were occasionally amended during the early stages of transition. However, those changes were only of a technical character. Since May 1993, the basket currencies have been reduced from five to just two currencies, the DEM and USD.

The shift from a very rigid system to a more flexible exchange rate arrangement was implemented in February 1996, by widening the fluctuation band to ±7.5%. From mid-1996 to March 1997, the CZK fluctuated in the stronger part of the band, with increasing deficits in trade and current accounts notwithstanding. The impact of capital account developments combined with the inertia of expectations was dominant. The trend was reversed in late April 1997 when the CZK shifted to a weaker part of the band. The deepening domestic imbalances, which were not followed by proper policy reaction, made investors nervous. The simultaneous economic turbulence in Thailand contributed to the spread of a foreign contagion and increased speculation against the CZK. The speculative attack culminated in the second half of May 1997, precipitating a currency crisis, and forced an exit from the pegged regime.

## Trade policy backsliding

In 1997, under the pressure of the currency crisis, a set of measures was introduced to cope with the unsustainable current account deficit and the accumulated imbalances. One component of this programme was the introduction of an import deposit on some consumer good imports. This import deposit was introduced as a temporary measure and was imposed on 21 April, 1997. Within a short period, it proved to be neither effective nor desirable. In August 1997, after a mere four months, the measure was phased out.

In the following period, the Czech authorities resorted to a few cases of product-specific trade policy interventions, mostly in the sphere of agricultural imports. Such examples were restrictions on apple imports from the EU, a specific tariff placed on wheat imports from Hungary and a proposed but never implemented restriction on pork imports from the EU. These selective interventions lacked a legal basis and the procedures were not consistent with its international commitments. As a policy, they were abolished within a short period and without any visible impact. Some were not made operational at all. One example of a viable intervention, consistent with existing legislation, was the imposition of an anti-dumping tariff on salt imports in October 2000.

# From recession to accelerating growth, from a current account balance to surging deficits in 1998–2001?

## Similarities and differences in the developments during 1998–2000 and 1996–97

After remaining low during the recession years of 1998 and 1999, the trade and current account deficits started to grow substantially in 2000 (cf. Table 7.9). A further increase in the deficit was expected in 2001. This development raised the concerns about what was going to emerge and whether it would be similar to the period between 1996 and 1997. The trade and current account deficits in the aforementioned period led to a currency crisis and, consequently, to some reversals in trade liberalization. The question to ask is whether there is a risk of repeated, unsustainable external imbalances that would call for trade policy interventions.

## Causal factors of current account imbalances
### *A built-in tendency towards trade and current account deficits*

Payment imbalances and external balance constraints are not confined to economies in transition or emerging market economies. However, they are more frequent in these types of economies. A 'built in' tendency towards current account deficits is related to the process of catching up. The acceleration of domestic growth was combined with the restructuring and modernizing production capacities and the increasing diversification of consumer

*Table 7.9* Czech Republic: balance of payments, 1998–2000 (in CZK bn)

| | 1998 | 1999 | 2000* |
|---|---|---|---|
| Current account | −43.1 | −54.2 | −91.4 |
| Trade balance | −82.4 | −65.8 | −126.8 |
| Service balance | 57.9 | 38.1 | 53.9 |
| Income balance | −31.7 | −44.2 | −29.5 |
| Transfers | 13.1 | 17.7 | 11.0 |
| Financial account | 94.3 | 106.6 | 129.6 |
| FDI | 115.9 | 215.7 | 172.8 |
| Portfolio investment | 34.5 | −48.3 | −68.2 |
| Other long term capital | −64.1 | −25.2 | −4.9 |
| Other short term capital | 8.1 | −35.7 | 31.7 |
| Errors and omissions | 11.3 | 4.9 | −6.4 |
| Change in reserves (- increase) | −62.6 | −57.1 | −31.6 |

*Source*: Czech National Bank.

*Notes*: *Preliminary data.

demand. These factors tend to result in trade and current account deficits. In the process, a higher export potential is realized after a notable time lag. This is due to increased imports of technology and production inputs. These causal links are relevant for the assessment of current account deficits in periods when acceleration of growth took place. This was the case in the mid-1990s and also in 2000–01.

### The impact of trade liberalization

The impact of trade liberalization on external payment balances may differ according to the time horizon. Through the increasing pressures of foreign competition, trade liberalization has contributed to the restructuring of domestic firms and to gains from the better allocation of resources. This improves the competitiveness of a country and works towards balanced development. Such effects, however, take time to materialize and occur over the medium run. In the short run, the potential to react to increasing demand and to meet foreign competition may be weakened. Such is the impact of sweeping liberalization, combined with implied uncertainty. This was the same issue present at the very start of transition process. The undervalued exchange rate was then looked upon as an instrument to avoid unsustainable external deficits.

### Cyclical developments, external shocks, and policy defects

The arguments for a built-in tendency towards trade and current account deficits, while catching up with the west and liberalizing trade in the aftermath of a centrally-planned economic regime, are specific to economies in transition. However, there are also other causes for external imbalances.

These include the standard ones linked to cyclical developments, external shocks and policy constraints.

The increasing weight of external flows made small transitional economies especially sensitive to cyclical development abroad and to external shocks. For the Czech Republic, as well as for other candidate countries, the EU markets are their 'anchor area'. This is having a dominating influence on domestic developments and external balances. This does not exclude potential contagions from developments elsewhere in the world economy. Again, the May 1997 currency crisis was at least partly triggered by an external contagion originating in East Asia.

Unlike in the former period, the impact of an economic slowdown in the EU area in growth and current account developments became a central issue in 2001. The Czech economy has shown resilience to external turmoil. This also seems to be the case for the marked deceleration of growth in the EU. Nevertheless, it cannot be immune to an adverse external environment, especially one in the EU. As a small, open economy with strong trade and investment links to the EU, the Czech Republic is susceptible to the conditions in its western neighbours. Germany, which absorbs 40 per cent of Czech exports and supplies more than one third of FDI is an example of this dependence. As of mid-2001, there were signs of slowing trade flows and of a negative impact on the trade balance. Though still in surplus, the trade balance with Germany deteriorated by CZK 21.8 bn in the first half of 2001 compared to same period in 2000.

## The characteristics of current account deficits, their ratios and sources of financing in 1994–97 and 1998–2000

In the mid-1990s, growing aggregate domestic demand was increasingly matched by foreign supply. The ratio of the current account deficit to GDP surged to a non-sustainable level of 7.4 per cent in 1996. This was still at 6.1 per cent in 1997, though the crisis broke up in May, and an induced adjustment followed (cf. Table 7.8).

In the period 2000–01, the expanding current account deficit (see Table 7.10) reflected particularly strong imports of capital equipment. This is related to the robust growth of FDI and fixed capital formation, as well as the impact of deteriorated terms of trade. This decline in the terms of trade was especially due to surging prices of imported fuels, oil and gas. Over time, the FDI inflows and strong imports of investment goods should boost competitiveness and export capacity. This shows that the economy is moving towards a balanced external payments system. Nevertheless, if a level of 5 per cent of GDP (ratio in the first quarter of 2001) persisted, the current account deficit would become a matter for concern. In addition, a deeper or more protracted slowdown in Europe and elsewhere in the world economy could result in a sharper than envisaged decline in export growth. However,

*Table 7.10*  Czech Republic: current account savings and investment ratios to GDP, 1998–2000

|  | 1998 | 1999 | 2000 |
|---|---|---|---|
| Current account / GDP | −2.3 | −2.9 | −4.6 |
| Domestic savings / GDP | 28.7 | 26.4 | 26.0 |
| Gross capital formation / GDP | 30.2 | 27.9 | 29.7 |
| Fixed capital formation / GDP | 29.0 | 27.9 | 28.3 |

*Source*:  Czech Statistical Office.

domestic growth, due to robust domestic demand, could accelerate for some time.

The financing of the current account deficit is more secure and stable in the early now than in 1994–97. The surplus in the financial account is dominated by FDI inflows, while short-term and portfolio flows contributed to outflows. In contrast, during the mid-1990s, volatile short-term and portfolio flows were more important (cf. Tables 7.5 and 7.9). Large, projected privatization revenues and greenfield investment projects are expected to provide sufficient financing in the forthcoming period as well. However, should privatization proceeds and FDI slow down (which is a realistic option for privatization deals), further expansion of the current account deficit and its financing may become a source of vulnerability.

Some other relevant indicators are the levels of external debt and the rate of its increase. A comparison of the share of short-term debt and its ratio to international reserves would also suggest a more favourable setting during the current stage in comparison to the previous period (Table 7.11, cf. Table 7.7). These indicators only confirm the recent experience, that the Czech Republic has become more resilient against external shocks than it was in the mid-1990s.

*Table 7.11*  Czech Republic: external debt and foreign exchange reserves, 1998–2000 (in USD bn, end of period)

|  | 1998 | 1999 | 2000 |
|---|---|---|---|
| External debt | | | |
| (convertible currencies) | 24.0 | 22.6 | 21.3 |
| long-term | 15.0 | 13.8 | 12.3 |
| short-term | 9.1 | 8.8 | 9.0 |
| (inconvertible currencies) | 0.3 | 0.2 | 0.2 |
| Official reserves (CNB) | 12.6 | 12.8 | 13.1 |

*Source*:  Czech National Bank.

## Factors influencing the sustainability of current account deficits

*Growth pattern*

After three years of negative growth, real GDP growth finally picked up to 2.9 per cent in 2000 (see Table 7.12). Available indicators show that even more robust growth will continue in 2001. This acceleration has been developing despite the marked slowdown in the EU and elsewhere in the world economy.

In qualitative terms, economic growth in the current period is different from that in the period before the 1997 turbulence. At that time, it was mainly driven by surging wages, which were in excess of labour productivity. At present, it is led by dynamic growth of investment, while consumption remains relatively moderate, and wage growth is lagging behind productivity increases.

*Microeconomic and institutional foundations of growth*

There were many factors that influenced the 1997 monetary crisis. These included a stalled privatization programme, serious problems in the banking sector, and defects in the area of creditor rights. The enforcement and implementation of bankruptcy proceedings were also important factors behind the evolving imbalances prior to the 1997 monetary crisis and the poor growth performance in the following period. Since then, there have been improvements in the legal framework and in the banking sector, which has undergone substantial changes. With the sale of Komerční Banka in 2001, all the major banks now have private foreign owners. Potential growth rates and economic performance should benefit from those changes.

*The fiscal situation*

Trends in the fiscal sphere have recently become a major concern. A more worrying trend was the medium-term public finance outlook. As a result of

*Table 7.12* Czech Republic: GDP growth and components of aggregate demand, 1998–2000 (y/y change in %, constant 1995 prices)

|  | **1998** | **1999** | **2000** |
|---|---|---|---|
| GDP | −1.2 | −0.4 | 2.9 |
| Household consumption | −2.2 | 2.3 | 1.8 |
| Public consumption | −2.4 | −0.1 | −1.3 |
| Gross fixed investment | 0.1 | −0.6 | 4.2 |
| Exports of goods | 11.6 | 8.4 | 19.9 |
| Services | 1.1 | −1.3 | 14.0 |
| Imports of goods | 6.2 | 5.1 | 20.7 |
| Services | 7.7 | 7.2 | 8.9 |

*Source*:   Czech Statistical Office.

resolving non-performing loans in the banking sector, public deficits and public debt will increase substantially. However, this would begin from a low level. In addition, weak economic performance in 1997–99 prompted the government to use fiscal policy as an instrument of demand stimulus. Even more constraining is the slow process of public expenditure reform. As a result, public spending continued to be too high even, in the mid-2000s. Accordingly, a simple way to cope with increasing fiscal deficits is to cut down on expenditures while trying not to increase revenues. The existence of a pro-cyclical demand stimulus in the changed environment could contribute further to increasing its current account deficit. The risk of twin deficits could then become quite tangible.

*Exchange rate developments and the exchange rate regime*

Once the financial system was liberalized the Czech economy grew in the first half of the 1990s. There was, however, an increasing inconsistency between the enhanced mobility of capital flows (reflected in massive inflows in 1994–96) and a rigid pegged exchange rate regime. This regime, while contributing to price and monetary stability in the previous stage, turned into a destabilising factor as conditions evolved. The advanced stage of financial openness proved to be inconsistent with a rigid exchange rate peg. The properties inherent to an economy in transition also persisted, in particular high inflation and interest rate differentials vis-à-vis the basket of currencies. The resulting bias in incentives stimulated massive flows of volatile capital. The pegged exchange rate, with a very narrow fluctuation band of ±0.5 per cent, could not cope with those incentives or with the destabilizing consequences of massive inflows and their reversals.

The increasing inconsistency between massive capital flows and the goals of monetary and price stability prompted the shift to a wider band in February 1996. This shift to a wider band of ±7.5 per cent increased the implied uncertainty for speculators and helped to deter short-term volatile capital flows. As a result, it temporarily decreased the destabilizing pressures. However, this was not a panacea to all the problems. It could not solve the deepening imbalances, nor could it substitute for a more adequate policy mix, which should have addressed them.

The process of nominal as well as real appreciation of the CZK continued throughout the existence of the Czech currency. Table 7.13 indicates the pace of the CZK's appreciation in terms of the effective exchange rate in the period between 1993 and 1997. The waves of capital inflows contributed to the persistence of appreciation, despite increasing deficits in the trade and current accounts. Since mid-1996, the CZK has fluctuated in the stronger part of the band, and its appreciation with respect to central parity exceeded 5 per cent in the first months of 1997. This was at a time when trade and current deficits were increasing.

*Table 7.13* Czech Republic: nominal and real effective exchange rates of the CZK, 1993–97 (in %, 1995=100)

|  |  | 1993 | 1994 | 1995 | 1996 | 1997 |
|---|---|---|---|---|---|---|
| Nominal Rate |  | 89.84 | 95.79 | 100.00 | 102.21 | 99.48 |
| Real Rate | PPI based | 100.58 | 99.94 | 100.00 | 103.03 | 102.15 |
|  | CPI based | 95.52 | 97.81 | 100.00 | 105.27 | 106.95 |

Source:  Czech National Bank.

*Note*:  > 100 = appreciation
       < 100 = depreciation

Table 7.14 gives respective data for the effective exchange rate developments in the post-crisis period. They suggest that after an upward shift in 1998, the level of real appreciation, in principle, stagnated.

Unlike the pegged rate, the floating exchange rate regime has provided a reasonable degree of flexibility and adjustment. Though subjected to the impact of capital flows which dominated in the short run, the regime is not as exposed to the waves of speculation.

### 'Good' and 'bad' current account imbalances?

Based on experience and lessons learned from the world economy, a few guidelines have been suggested as indicators of external deficit viability. The exceeding of these indicators should send warning signals. However, their reading is not a simple matter. It is not a single factor that decides, but a range of factors that determine where the limits of a sustainable external deficit may be. Each case is also likely to be different. As a general rule, the incurred deficits should not exceed the level viewed by foreign investors as safe and manageable. This requires avoiding a negative change in their sentiments and the resulting capital outflow. Experience shows that the line between a current account deficit viewed by markets as benign and sustainable, and a current account deficit that could provoke an adverse market reaction may be very thin.

*Table 7.14* Czech Republic: nominal and real effective exchange rates of the CZK, 1998–2000 (in %, 1995=100)

|  |  | 1998 | 1999 | 2000 |
|---|---|---|---|---|
| Nominal Rate |  | 102.47 | 106.81 | 106.79 |
| Real Rate | PPI based | 109.41 | 111.66 | 109.34 |
|  | CPI based | 117.27 | 117.33 | 116.11 |

*Source*:  Czech National Bank.

*Note*:  >100 = appreciation.

Consequently, the sought after distinction between 'good' and 'bad' current account imbalances is quite ambiguous. One of the applied propositions is that a current account deficit, due to increased investment, is better than one financing consumption growth. The related argument points to the potential for repayment. The higher the generated potential of future export proceeds is, the more likely the self-extinguishing character of current deficits can be. If such a generated potential is high enough, the issue is only a time lag between current investment financing and future debt repayment.

Another proposition is inferred from the so-called Lawson doctrine. It claims that current account deficits should not be such a matter for concern when they reflect the discrepancy between private savings and private investment. This is because such a gap should be self-correcting. The situation is different when the current account deficit is primarily due to public sector activities. There are, however, powerful objections to the validity of the Lawson doctrine on the basis of both experience and theoretical reasoning.

Nevertheless, if we were to assess the similarities and differences between current account deficits in the period 1996–97 and in 2000–01 for distinctions between 'bad' and 'good' current account deficits, then most features in the former period signalled a 'bad' deficit. In contrast, the 'good' properties prevailed in the latter period. It can, therefore, be inferred that the risk of trade policy backsliding and reversals due to macroeconomic constraints is relatively low in the future.

## Summary: factors of trade policy instability

There are many factors that may initiate pressure towards policy reversals and the temptation to substitute a desirable adjustment elsewhere in the economy. This is mainly through trade protection measures. These usually consist of two types:

- The first type results from macroeconomic developments and policies, in particular due to external balance constraint and payment imbalances. The assessment of these causal factors and of related arguments has been the subject of this study.
- The other group of factors reflects a variety of branch and commodity-specific, regional, environmental, health, security and safety considerations and standards.

The experience from Czech economic transition during the 1990s leads to a few conclusions on the trade policy regime and its potential instability:

(a) There have been sweeping structural and institutional changes as a result of the transition process. There have also been external as well as 'domestically produced' shocks and turbulence. However, in spite of all these adverse conditions, a liberal trade regime has been maintained since the very beginning of the transition period. Trade policy instruments have been kept relatively stable, market-based and transparent throughout the decade. This is a remarkable achievement for an economy in transition and when compared with the experience of some neighbouring countries.

(b) The liberal trade policy stance has been paralleled by advancing capital flow liberalization. This confirms that, over the 1990s, the preference towards a free trade environment has prevailed. The Czech authorities have shown that they were committed to complying with their international obligations.

(c) Nevertheless, on a few occasions, some slippages and policy reversals did occur. There has been the temptation to resort to various forms of trade protection.

(d) The Czech development process and the experience of other countries has shown us that the implementation of trade policy instruments can arise when trade and current account deficits are heading towards unsustainable levels. In such a situation, the authorities are confronted with the risk of losing of control over further developments. Alternatively, trade policy restrictions can be imposed when a currency and balance of payment crisis has already occurred. Being a consequence of the crisis, they are considered to be an instrument for its resolution. This was the case of the Czech currency crisis in 1997.

(e) The occurrence of trade and current account deficits need not be negative or detrimental. Following from the identities of national accounting, the current account deficit enables circumvention of the limit of national savings. Therefore, the volume of investment can be increased beyond the level of national savings. There are, however, limits to the viability of external deficits. When exceeded, the national economy is put under pressure and faces the risk of external liquidity or a solvency crisis. At this point, the arguments for trade policy slippages are raised.

(f) The sustainability of trade and current account deficits is related to the developments in the financial account. Capital inflows provide financing to current account deficits. Therefore, it matters whether they are stable or volatile, and whether there are substantial reversals of inflows. In addition, why these reversals occur is also important

In the course of the 1990s, the role of capital flows in the world economy has substantially increased. This was particularly due to advancing globalization, liberalization and technological development. The benefits of a much higher volume of capital flows, and their increased mobility, for economic development were accompanied by the increasing vulnerability of national economies. These were due to massive capital flow reversals and volatility. The implied risk applies in particular to small economies in transition, given their underdeveloped and fragile financial infrastructure.

(g)  To assess capital flow liberalization, its merits and costs, in the case of the Czech Republic is not an easy task. The ongoing discussion on the subject, in relation to emerging market economies, has raised a number of arguments in favour and against.

There are few major disagreements with the commitment to capital flow liberalization. However, the speed, sequencing, and relationship to the progress of liberalization in other fields and policies, including the exchange rate regime, are subject to wider discussion. With the benefit of hindsight, it can be argued that there have been many constraining factors. These include an underdeveloped financial infrastructure, defects in the existing legal system, and barriers to the enforcement of law. There were also limits on the effectiveness of macroeconomic policies. These could be seen during the liberalizing capital flows of the mid-1990s. Due to these circumstances, the side effects of capital flows and their consequences may have contributed to increased vulnerability of the Czech economy and to the resulting reversals (or, at least, attempts at such reversals) in trade policies.

This conclusion does not imply, however, the desirability of capital controls in the conditions of the Czech economy. Reaping the benefits of capital market access while coping safely with the risk of capital flow volatility requires a sound and efficient domestic financial infrastructure. It also requires proper coordination with other policies as well as sound microeconomic and institutional foundations of macroeconomic stability and growth.

(h)  In coping with external deficits and consequences of massive capital flows, with their reversals and volatility, several lines of adjustment and 'defence' can be applied. Adjustment of the macroeconomic policy setting, structural changes, and institutional reforms enhancing the flexibility and efficiency of domestic markets represent homeward oriented options. An alternative way is adjustment through the exchange rate mechanism. Trade policy reversals, backsliding from the achieved level of liberalization, as well as parallel reversals in the area of capital flows represent attempts to 'substitute' for the mentioned methods of adjustment.

## Notes

1  The 3-month PRIBOR minus the weighted average of USD (LIBOR) and DM (FIBOR) moved in the range of 6–8.5% in 1995–96.

2  Thanks to the significant, gradual upgrading in the ratings of sovereign and private debt, those credits were cheaper than domestically provided.

# Index

Printed in the United States
By Bookmasters